Special thanks to:
Professor Tobias Gibson
Professor Lloyd Fitzpatrick
Editor Sheryl Johnson
And all the many sources researched for their research help and advice in the making of this book.

TABLE OF CONTENTS

Chapter 1: The W's and H of Your Life   9
Chapter 2: Life is Either an Interview or an Ambush 19
Chapter 3: The WHO Question and Confrontations   33
   Who are you...really?
   Who do you really think you will be fighting?
   Who are you legally?
   Who is around you?
   Who is trained to fight?
   Who will judge your actions?
   Who (or what) is suspicious?
   Who will take avenge, revenge, vengeance?
   Who will teach you?
   Who inspires you?
   Who will stand beside you?

Chapter 4: The WHAT Questions & Confrontations   79
   What is the big picture of safety?
   What if? What happens next continuum?
   What makes you a target? A potential victim?
   What is winning anyway?
   What then is an orderly retreat?
   What is a citizen's arrest?
   What clothes do you train in?
   What training should you prioritize?
   What can you do and what shouldn't you do.
   What's it gonna' take...game.
   What would Jesus do?
   What is the right reason?
   What is your perception of your next fight
   What (if any) weapons?

What is "weapon brandishing?"
What happens to your body in a confrontation?
What should I do alone, having a heart attack?

Chapter 5: The Where Question & Confrontations 145
Where could you have problem
Where do you live?
Where are you, when away?
Where do you work?
Where can you sit, stand, run, hide or fight?
Where do the fight "collisions," occur?
Are parking lots really most dangerous places?
Where should you shoot, hit, stab or slash an enemy?
Where is the planned, second location?
Where is Urban Combatives useful?
Where do crimes occur?
Where are the cameras?
Falling where?
Where should I get to? Where should I be?
Where exactly should I look when in a fight?

Chapter 6: The WHEN Question & Confrontations 221
When are you the weakest in your decision ma
When are the dangerous times you drive?
When will I be attacked-involved in crime
When - the before, the during, the after.
When must I be alert ?
When are you the weakest in your decision making?
The 3 managements - anger, fear and pain.
When are the most dangerous for divorced women
When will active-mass shooters probably strike?

CONFRONTATIONS!
WHO, WHAT, WHERE, WHEN, HOW AND WHY
Winning and/or Surviving Through the
Argument and the Ambush

Copyright 2026
Published by High Home Endeavors: Books!

Other Books by W. Hock Hochheim
Fightin' Words
Dead Right There
Don't Even Think About It!
Rust in Pieces
Gunther 1: Gunther! The Law West of Medieval
Gunther 2: Guns of the China Alamo
Gunther 3: Last of the Gunslingers
Gunther 4: Riders of the Khyber Pass
Gunther 5: Rio Grande Black Magic
Gunther 6: The Horse Killers
Kellog 1: Kill Them Back!
Kellog 2: Face the Muzak
Kellog 3: Takedown the Take
Impact Weapon Combatives
Knife Combatives
Training Mission Series
The Great Escapes of Pancho Villa
Renegade General 1: Swellen's Reckoning
Renegade General 2: Swellen's Orphans
Migraine! The Ws and the H

Chapter 7: The HOW Question & Confrontations 257
  How will they approach you?
  How long before your perishable skills perish?
  How fast is my reaction time?
  How do you control yourself? Control adrenaline?
  How many counters are there?
  How do fights physically start?
  How complicated could, should fighting be?
  How to handle the argument confrontation"
  How do you detect and avoid an ambush?
  How do you mentally survive critical incidents
  How small is your mind?
  How long is a martial second?
  How can I perform better?
  How do I make my office workplace safer?
  How can I de-escalate an angry attacker?
  How should I stand when I am confronted?

Chapter 8: The WHY Question & Confrontations 339
  Why the evil that men (and women) do?
  Why go there and why are you still there?
  Why do some survive, and some don't?
  Why then should I fight? Lines in the sand.

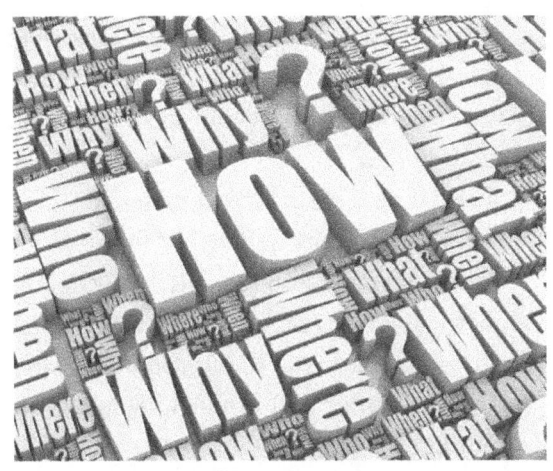

Chapter 1:
The Ws and H In Your Life and Dissecting Confrontations.

"Who, What, Where, When, How, and Why." I first encountered these fundamental questions in the early 1970s while attending the U.S. Army Military Police Academy. They served as a checklist for police officers to collect information and write clear, thorough reports. Answering these questions provided both broad and detailed insights. However, I later learned that a detective must go even deeper, uncovering layers of meaning. A prosecutor then takes this investigation further, scrutinizing every detail. By the time a case reaches trial, the smallest, most unexpected piece of information can become critically important.

Throughout this process, one develops skills in research, debate, and argumentation. These questions also form the foundation of journalism, often referred to as the "5 Ws and 1 H." If a journalist can effectively answer these six questions, they can present a comprehensive and accurate account of an event or topic.

Over time, I realized that these questions are not just useful for report writing but for all aspects of decision-making and life itself. This framework applies to everything, from purchasing a home or car to choosing a spouse, raising children, selecting medical insurance, or even picking a pet.

While these questions serve as a valuable tool for navigating life, this book focuses on their application in self-defense, training, protection, safety, and survival. Whether you are using an ATM safely, protecting a high-profile individual, or even planning a D-Day invasion, answering these questions effectively can make all the difference.

Having served as a bodyguard for numerous famous individuals and worked with the U.S. Secret Service on two presidential details, I have relied on these questions to guide my decisions. Whether working solo or as part of a team, these principles have shaped my approach to security and preparedness. Since the early 1990s, I have emphasized their importance in all my training courses. They serve as a foundation for conflict forecasting. They require attention to key aspects:

- Duality. "This isn't just about you!" You've probably heard this before. Every W and H question applies not just to you but also to your opponent (or multiple opponents), even

the total situation at hand. Understanding duality means recognizing the perspectives and actions of others in any given scenario.
- Macro and Micro Analysis.
    - Macro: The big-picture questions and answers.
    - Micro: The finer details that provide clarity.
    - Both perspectives must be examined to form a complete understanding.

- Endless Rounds of Inquiry? Endless sounds intimidating. It just means "don't quit asking." Answering one question often leads to revisiting others. Each answer provides new insights, requiring continual reassessment. Within a single category, ask all six questions repeatedly to refine your understanding. This approach will become clearer as you progress through this book. Each chapter concludes with a review box to remind-reinforce these principles. I am not sure your investigations ever come to complete end. Sometimes. Perhaps.

Setting the Stage for Critical Thinking. Mental preparation and crisis rehearsal are essential. Answering these six questions helps you "set the stage" for forecasting and decision-making. Just as a director carefully sets a scene in a movie, you must envision potential scenarios and prepare accordingly.

However, research and preparation are only as good as the accuracy of the conclusions drawn. In the next section, we will explore critical thinking, forecasting, and preparation. How well can you anticipate and plan for verbal or physical confrontations? Can you set the stage for discussions, arguments, conflicts, crimes, or even war? Ultimately, you are the producer, director, and star of your own story. The plot thickens…

Decision Making: Answering the W's and H's. Decision-making and forecasting, - what Daniel Kahneman famously described as thinking "fast and slow, "are essential skills in navigating the real world. Decisions range from deliberate and slow to split-second choices made under pressure. Each comes with consequences, but the fastest ones? Those made in the heat of mental and physical ambushes. They must be lightning fast. The book shown is a must-read.

Analysis paralysis? There's much to consider, but please, don't become Hamlet, lost in the Shakespearean cosmos of indecision. Hamlet couldn't make a choice (which, of course, made for a much longer play). Sometimes, decision time is critical.

This book however might be called, *"Deciding Fast and Slow!"* How much time do you have in verbal and physical confrontations? That depends on the situation, but let's not linger past the our "expiration date." Here are the common types of decision-making, as debated by experts in the field:

- *Calculated Decisions* – When time allows for careful consideration. Gather information, analyze possibilities and decide. Straight-forward, right?
- Intuitive or Gut Decisions – Rooted in knowledge and experience, these decisions emerge when there's no time for deliberate analysis. Recognizing patterns from past learning allows for immediate action. Many experts argue that trusting your educated gut is often the right move. But can you always trust your education, the very source of intuitive, gut decisions? Even these same experts admit that, at times, gut instincts can be wrong.
- *Heuristics* - relying on patterns, is a category that is often presented and debated within the confines of decision making and intuition. "Explore Psychology" (and many other sources) report, "In psychology, a heuristic is a mental shortcut or rule of thumb that helps people make decisions quickly and efficiently, even when they don't have all the relevant information. Heuristics are generalizations that can reduce cognitive load

and help people use reason and past experience to solve problems. They can be effective for making immediate judgments, but they can also lead to irrational or inaccurate conclusions.
- *Hick's Law versus George Miller's Chunking Law.* Time. Response. Decision making. We will get into that deeply in an upcoming segment. In some verbal and certainly some physical confrontations time is "of the essence."
- What "hat" are you wearing when making decisions? The truth is, you're slightly different people at different times when questioning, when answering, when making decisions. You're not the same person at home with your kids as you are at work, coaching a team, attending church, or standing on a battlefield. Your mission, and your solutions, change depending on the who, what, where, when, how, and why of your moment and your mindset. So, ignorance alone isn't the only challenge your quality of thinking in context may also be in question. This is the essence of the "duality" in thinking: the tension between your rational mind and your biased mind.

If you see something, say (or do) something. We've all heard the phrase, "If you see something, say something." It's catchy, post-9/11 government advice. But...see what? Say to whom? And why, exactly? Does something need to be done...now! The great "They" have never really explained that part. This book will define some of it for you. Are you thinking or doing fast or slow?

The Ws and H are tools to help you gather reliable information. They build pattern recognition and develop intuitive heuristics, what some call forecasting. We'll dissect these questions, slice and dice them, research, experiment, dig deep, break things apart and put them back together.

You'll never know it all, nobody does, but you can move closer to understanding. "Fortune favors the prepared (mind)" is the old saying.

Prep probabilities versus possibilities. We'll never ask or answer every possibility. Seek probabilities first. Possibilities next, as they are quite endless. This book is a kick-starter for citizens, police and the military to spark and inspire your mind.

"Understanding a question is half an answer."
"The depth of a question reveals the depth of your mind."   - Socrates

"A problem well stated is a problem half-solved."
-Charles Kettrering

"If I had an hour to solve a problem I'd spend 55 minutes thinking about the problem and 5 minutes thinking about solutions."

"The important thing is not to stop questioning. Curiosity has its own reason for existing."
- Albert Einstein

"A wise man's question contains half the answer."
-Solomon Ibn Gabirol

"Judge a man by his questions rather than by his answers." - Voltaire

"The power to question is the basis of all human progress." -Indira Gandhi

"The master key of knowledge is, indeed, a persistent and frequent questioning." – Peter Abelard

"Our research reveals that strategic questions can be grouped into five domains. Investigative, speculative, productive, interpretive and subjective. Each question unlocks a different aspect of the decision-making process."
- Chevallier, Dalsace, Barsoux, Harvard Business Review

"Leadership involves getting everybody to ask and answer questions." - Jensen Huang

The Ws and H, questioning, is essential in life. And we will use it this time to dissect the art, the way, the means, meanings of "confrontation." The best definition I could find for the noun is "facing a challenging situation or person directly, often an open conflict of ideas or physical force, while War is a large-scale, armed struggle-confrontation between nations or groups, characterized by organized, violent warfare and deep political hostility. Confrontation can be personal (a tough talk) or societal (protests), but war is inherently violent, involving military force to achieve incompatible political goals."

Do understand that this book, albeit with some stage-setting about verbal interviews and arguments, is mostly about worst case scenarios where crime, war and violence occur.

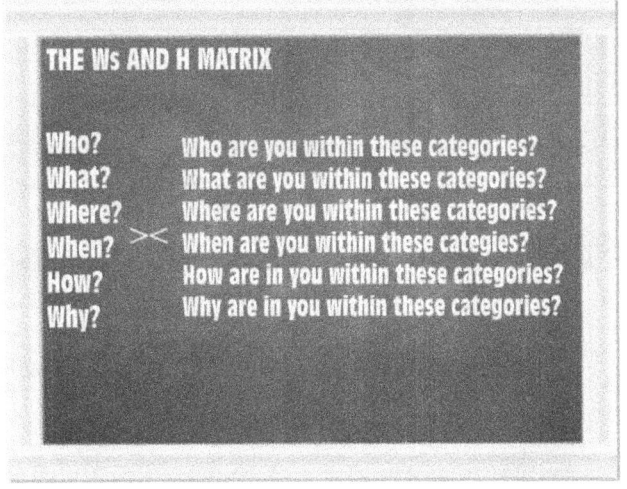

THE Ws AND H MATRIX

Who?   Who are you within these categories?
What?   What are you within these categories?
Where?   Where are you within these categories?
When?   When are you within these categies?
How?   How are in you within these categories?
Why?   Why are in you within these categories?

Chapter 2: Life is Either an Interview or an Ambush Confrontation

The "Scripts of Life." Some are confrontations. Some not. I have nicknamed the dialogue from chit chats on up through wars as "the scripts of life." Each category has its scripts, words, and lingo people repeat. Ways people talk. Yes, they can be novel at times, but they often follow certain scripts, rabbit holes and funnels-tunnels. So many of the exchanges in life are typical, you might say virtually pre-rehearsed in a way. It is actually quite handy to view banter, conversations, arguments and confrontations with the common scripts of life in mind. If you take notice, you know genuine conversation from rehearsed conversation.

When I declare that "life is either an interview or an ambush" here or in seminars, I mean that there are ambushes inside small and big interviews, small and big arguments, small and big fights, small and big crimes, small and big wars. We will start from scratch to provide a basis.

The Chit-chat. We also must remember that each day around the world there are billions of successful, safe, friendly-cordial, verbal interactions that are quick, shallow, "chit chats" between scores of peoples of all races, creeds and colors. This massive positive reality creates a cooperative culture, and a healthy society. This good news will never really be publicized within the headline seeking news media and political groups that wear you down with shock and fear for ratings. "Sensei Webster" defines a "chit chat" as "to talk informally about matters that are not really important at all."

With many people in life, one stays within the chit chat range for a variety of reasons. No time. Not interested. Whatever. Deep down, you know and feel this impersonal chat range when engaged in them. These sessions are with strangers or acquaintances. The stranger/acquaintance chit chat might develop into a question-and-answer interview of sorts. Buying a cup of coffee is a chit-chat, but casually bring up politics or sports in the process of buying coffee? And trouble might also be brewing. Your little chit-chat might became a give-and-take interview, or then - an argument. Then...

The interview. While chit-chat is usually brief and informal, interviews can range from short to

extensive. "Professor Google" defines an interview as:
1. "A meeting of people face-to-face, especially for an interactive consultation." (I like that word, "consultation.").
2. "A structured or unstructured conversation where one participant asks questions, and the other provides answers."
3. "To become involved intentionally or unintentionally in a difficult situation in order to change it or improve it or prevent it from getting worse."

We are probably most familiar with the term "interview" in the context of job interviews or media interviews, but in reality, we interview each other in many ways. The other person or people respond with statements and/or questions. Like chit-chat, interviews can take place between strangers or acquaintances, and for some, an interview might become…testy. Challenging. Even aggravating. "Testy" can sometimes escalate into an argument.

Arguments. Our dialogue grandmasters define an argument as: "An exchange of diverging or opposite views, typically a heated or angry one." An argument might be considered a more intense version of an interview. In rare cases, it can escalate further, leading to physical altercations between strangers, spouses, friends, or relatives.

Yes, a fight! And once physical contact occurs, it may be classified as a crime or in the biggest of pictures, an act of war. After all, in most countries, laying hands on someone, or using weapons such as sticks, knives, bullets, or bombs, is considered a crime of varying degrees or, in big cases, a military engagement, conflict or war.

Ambush. Within every chit-chat, interview, argument, fight, crime, or war, there can also be… an ambush, a surprise attack of some sort. A thesaurus of words surrounds "ambush," helping to define it for us:

- "bushwhack, decoy, ensnare, entrap, mugging, hide, hook, jump, lurk, net, pounce upon, infiltrate, invade. An ambush might come in the form of a verbal "gotcha" question or an unexpected comment, catching someone off guard. Or yes, it could be a physical attack.

- As "General Wikipedia" explains: "A military ambush is a surprise attack carried out by people lying in wait in a concealed position. The concealed position itself or the concealed person(s) may also be called an 'ambush.
- Ambushes, as a basic fighting tactic of soldiers or criminals, have been used consistently throughout history, from ancient to modern interactions, business and warfare."

A verbal ambush. A verbal interview-confrontation often includes a surprise, ambush "Gotcha' question. Such a question is a query or statement, often posed by friends, spouses, co-workers, customer service, etc., but we see mostly journalists or lawyers. Lawyers in court or depositions layout the trap/ Building…building…it's designed to trap or trick the person being questioned into making a statement that is damaging, discreditable, or embarrassing. The questioner is often not seeking an honest answer but rather an outcome that can be used against the interviewee. Key characteristics of a gotcha question:

- Manipulative Intent: The primary goal is to make the interviewee look foolish or inconsistent, rather than to seek information or understanding.

- Loaded or Leading Language: The question often includes a presupposition or a false premise, making it difficult to answer without implicitly validating the negative assumption. The classic example is, "When did you stop beating your wife?".
- Lack of a "Good" Answer: Any simple yes or no answer will put the person in a bad light, forcing them into a defensive position.
- Context Manipulation: They can be used to lead a person into a sensitive area and then confront them with information designed to contradict or discredit their position, often in a time-constrained format like a sound bite, where nuance is impossible to convey.
- Versatile Use: While commonly associated with political interviews and media, these tactics can also appear in personal relationships, workplaces, and social media interactions.
- The term "gotcha" derives from the exclamation "I've got you" or "I've caught you," referring to the moment the person falls into the trap.

Adventures in talking... Here are two interview-only, verbal ambush examples in common citizen life.

Example 1: Interview-
A business owner asks a plumber to inspect his building. The plumber says, "I think these water pipes in the ceiling will burst in about a month. They are very old." They engage in calm problem-solving give-and-take interview.

Example 1: Ambush-
At the other extreme, like an ambush version of similar events, the owner fails to order the early inspection, strolls into his business one morning to find his employees soaked, water raining from the above pipes. The workers race up to his face, demanding, "Something must be done! If we don't work today will we lose the company!"

This is a lifestyle-ambush decision of sorts. The owner is under surprise stress to quickly solve a problem. His day, his business has indeed been ambushed by circumstances… life.

Example 2: Interview
A common verbal ambush example, a husband has done something wrong. He has sent his mother $3,000 without telling his wife. And his wife discovers it. He walks through the front door happy whistling, and the wife says calmly, "Jerry, can we sit down

and talk about something? Can you explain your thinking about this money thing with your mother?

Example 2: Ambush -
Jerry walks in. wife charges him in anger yelling, "Did you send your mother $3,000? (Or any domestic topic.) Surprised and unprepared, verbally "ambushed" what will he say? His actions may be reasonable and justified once calmly explained, but in this ambush-mode he might babble, be confused, making matters worse.

Here are more examples of more talking adventures and how they might escalate into a confrontation.

	"Your teeth are very yellow. You should use Exogene toothpaste
Chat version	"Oh? Yeah. I will look into that," you say, despite your contrary opinion about that toothpaste and its ingredients, you just let it go.
	"Your teeth are very yellow. You should use Exogene toothpaste."
	"You think? Are you nuts! That crap causes cancer!"
Argument?	"No, it doesn't! Who says?"
	"Yes, it does!"

It's often easier to take a conversational approach, even if you disagree about Exogene toothpaste. However, some people might respond emotionally. Perhaps insulted? Do you, would you feel insulted because:

- Someone suggests your teeth are yellow?
- Someone assumes they are smarter than you about something?
- Someone fails to recognize the dangers of cancer-causing products, something you're passionate about. You want to educate them?

In the world of policing, this is still an "interview." It would be wise for citizens to adopt this professional calm in many situations until there's physical contact. By the way, take it from me, it's very hard.

This "interview" percolated into an argument, then a fight, interrupted.

Here you see a common street crime ambush, a mugging. These attacks typically happen fast and aggressively. Attacks from the rear or with assailants yelling, threatening, and shoving to create confusion, fear and submission.

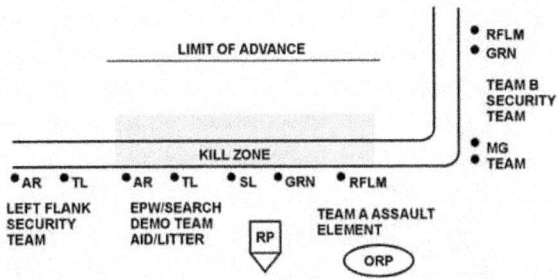

Above, a military ambush example. The classic L-Shaped Ambush Formation The L-Shaped Ambush is perfect for many military style surprise attacks with two or more lines of fire. There are numerous configurations.

The surprise attack has defeated some of the greatest militaries in history, left crime victims hurt or dead, and even turned the tide in both indefensible and righteous arguments. On ambushes, the often-quoted Chinese strategist Sun Tzu advised:
> "In conflict, direct confrontation will lead to engagement, and surprise will lead to victory. Those who are skilled in producing surprises will win. Such tacticians are as versatile as the changes in heaven and earth."

So, how do you prepare for the element of surprise in both common crimes and warfare? This book aims to help. My best way to help you organize your preparation, for both the interviews and ambushes of life, is through the Ws and H questions. Consider each question carefully.

> **Interview and Ambush Summary.**
> These events do not have to appear in order. Here is a kick-off, review-reference list.
>
> 1: Chit chat = non confrontational talk.
>
> 2: Interviews = People interacting beyond chit chat. Question and answer.
>
> 3: Arguments = People interacting with some heat. This still should be considered an interview. If it looks like a fight will erupt, this period too is still very much an interview, and very much an assessment until there's contact.
>
> 4: Ambush = a surprise verbal or physical attack.
>
> 5: Physical Fighting = Things percolates to physical violence. Hand, and/or stick, knife, gun, or other tools and objects.

A Note on My Approach: As we progress through this book, you won't find step-by-step physical techniques for hand-to-hand, stick, knife, or firearm physical defenses. I cover those in other books with thousands of instructional photos. This book is all about thinking-preparing, forecasting such confrontations. Each question is a thought experiment designed to sharpen your mindset. Due to the nature of each question, topics might seem split at times, but you must read the whole book! It will all be covered for you in a logical question and answer format.

Chapter 3:
The WHO Questions in Confrontations

"Our sense of our own self-worth and our own self-confidence is derived from judgments about our peer group. So, if you put someone in a very, very highly competitive pond, they are going to reach very different conclusions about who they are and what they are capable of than if you put them in a less selective pond, a smaller pond." - Malcolm Gladwell

WHO Question 1: Who Are You… Really? Who are you, deep down? Your job? Your physical condition? Your age? Your strengths? Your weaknesses? Your mind? How do you see yourself, real, not real, or somewhere in between? How much endurance do you have, how much gas? How much explosive power, how much TNT? Are you ready for a give-and take interview or argument? The sudden impact of an ambush fight? Are you bigger, faster, and stronger than you think? Let's look at the young vs. The old.

The young. Some young people overestimate their abilities, falling victim to what's known as the Dunning-Kruger Effect, a psychological phenomenon identified by Cornell University researchers David Dunning and Justin Kruger. This lack of experience effect explains why some individuals believe they are more skilled than they actually are. The causes include: Lack of self-awareness, they don't recognize their own incompetence, leading to an inflated self-perception.

The idea that the brain fully matures at or around 25 has been around for a while, but it has roots in neuroscience. The prefrontal cortex, crucial for judgment, impulse control, and planning, is one of the last areas to finish developing, with significant maturation extending into the mid-20s and even later. While many cognitive functions are present earlier, this refinement leads to better emotional regulation, focus, and decision-making into the late 20s and beyond, making 25 a rough guideline and average for significant functional maturity. Look around and you can see this development in societies.

Overconfidence despite lack of will, experience and skill. Youth often assume "they're better and smarter than they are." Sometimes they are. Sometimes they're not.

The old. But what about the opposite? Some older people underestimate themselves, despite being more capable than they believe. This can be linked to Imposter Syndrome (or Imposter Phenomenon), a term coined in 1978 by psychologists Suzanne Imes and Pauline Rose Clance. This condition causes people to doubt their skills, intelligence, or accomplishments, even in the face of evidence. I don't like the name of it – imposter - it is misleading, but it is what it is.

A study found that as many as 82% of people have felt like a fraud at some point, including Albert Einstein. Those with imposter syndrome may feel undeserving of their success, despite having proof like awards, degrees, or trophies. One key factor? As people age, they may increasingly doubt their abilities.

When older individuals believe they are less capable or less valuable due to age, it's often self-directed ageism, internalizing negative stereotypes about aging.

This mindset can reduce self-worth, hinder performance, and create the feeling that one's best years are behind them. While aging inevitably brings change, the real question is: To what extent? And when?

Wise individuals in the martial, law enforcement, and military worlds understand the importance of curating and maintaining small sets of favored discussion and fighting tactics and techniques, ones that align with their experiences, size, strength, abilities, and age. Wise, martial arts legend Dan Inosanto once advised in a seminar I attended, that every five years or so, it's wise to reassess your "favorites list." Can you still execute these techniques effectively? Do you need to adapt and replace them
with something better suited to your current abilities and age?

Fans and boxing experts recognized that Father Time did slow even the great Mike Tyson down in the 2024 fight versus Jake Paul.

I know that one day, despite all my effort in maintaining strength, endurance, and skill, my primary means of self-defense may simply be a hammerless, snub-nose revolver tucked into my coat pocket. But when will that day come? The reality of perishable skills is something we'll explore later in this book.

<u>*Who are you mentally and physically for arguments, self-defense and survival fighting?*</u> Here's a breakdown of key considerations for preparation:

- Mental and Physical Sharpness – Is your mind sharp from mental exercises, rest and good health? For discussions, interviews and arguments, are you abreast of the unbiased facts?
- Is your body sharp from regular exercise and conditioning.

- Size and Strength – How big are you? Your team? Your potential opponents? Consider your height and weight, are they an asset or a disadvantage? Are you shorter, average, or taller than most? Prepare to fight with and against the averages. (Bruce lee knew he was smaller and thinner than most, so became faster and stronger than most.
- Reach and Arm Length – In boxing and kick boxing, reach is a critical factor. How do your measure up?
- Gas and Dynamite – How much endurance (gas) do you have? How much explosive power (dynamite) can you generate? Know your limits and maximize your strengths.
- Fakes, Feints, and Setups – Do you have reliable setups you trust? Can you execute effective three-to-four-step combinations with confidence?
- Very few people are trained to fight. Decade after decade, business studies show that few people are trained to fight. Even most people in the military are not trained for close-quarters fighting. So be prepared for the "wild, angry man." (More on this later.)
- Self-Assessment: Where do you truly stand? From youth to old age, from unskilled to highly trained, every stage comes with its own realities. Where do you really fit in terms of ability and expectations?
- Honest self-assessment through training, real-world experience, and answering these key

questions can bring clarity. Your mentality, physicality, and life experiences all contribute to your effectiveness. Ultimately, your endgame is survival, with no arrests, no lawsuits, and no hospital visits.
- Stay sharp, stay prepared, and stay honest with yourself.

"Who are we, who is each one of us, if not a combination of experiences, information, stories we have read and seen, things imagined? Each life is an encyclopedia, a library, an inventory of objects, a series of styles, and everything can be constantly shuffled and reordered in every way conceivable."
- Italo Calvino, Six Memos

WHO Question 2: Who do you really think you will be discussing things, arguing with and fighting with?
We've just asked, "Who are you?" in question 1, and now, with the duality mandate, we must ask, who do you really think you will be messing with? Consider your geography, politics, lifestyle, the times, crime trends, and war trends.

<u>*Who are you arguing with anyway?*</u> In arguments, there is a "why bother" arguing factor. You might assess their possible education, the political silo-bubble they live in, (that bubble's pretty hard one to pop), background and I.Q. I have learned to avoid touchy discussions with numerous people. Why should I engage? I am not keeping score or running for office.

I find that most people cannot go "5 questions deep" on any subject, without getting angry or emotional. This can lead to actual shouting, insults and fights. In short, losing their unbiased way.

> "Irrational is a strong word which connotes impulsivity, emotionality and a stubborn resistance to reasonable argument."
> – Danial Kahneman

In fairness, rational or irrational, it is hard for all of us to recall, have handy on the very tips of out tongues, all the facts and figures that professional experts know about any subject. Cut people and yourself some slack on this when "ambush debating."

Another "who" point. I am not a big fan of promoting the overall concept of I.Q. tests and rates. Researchers emphasize that differences in average IQ scores across countries can reflect socio-economic factors, education access, colloquialism, health, nutrition, and cultural biases in test design, rather than innate mental ability differences. (In other words, try to complete a crossword puzzle in another country.) But many scientists urge caution when interpreting or ranking nations by IQ because of these limitations. But it is *generally* accepted in the USA that the score of 80-89 is considered below average, which is said to be some 16 % of the population or 53 million people. A score of 70-79 is considered borderline low, some 7 percent of the population or 22 million. Very low? Below 70, or 2.2%, or some 7 million people. Just saying - this is a total of 25% of the population. You want to argue nuances with some of these people? You want to fight these people when push comes to shove? (Let's not even begin with emotional intelligence.)

Then of course, you've also heard the expression, "He's so smart he's stupid." There's that.

<u>Who must we worry about? Fear? Fight?</u> Every hand-to-hand, stick, knife, or gunfight is a highly situational trauma and drama with unique circumstances and consequences. Staying informed about crime trends and antisocial behaviors through news sources is crucial in identifying who you might actually be fighting, may be a "friend" or foe? Unfortunately, most martial arts schools or seminars fail to address any of these realities. They often train against opponents who are essentially mirror images of themselves. Boxers fight boxers. Wrestlers fight wrestlers. Etc.

Other martial training programs make the mistake of treating every opponent as if they were a Nazi commando, plucking out eyeballs, knocking them down and out, and crushing their throat and face with multiple boot stomps. Unless you are in a war, this level of force is excessive in many situations and could land you in jail.

So, who do we fight? We can categorize opponents into three generalized groups.
- Group 1: People you know, know well or acquaintances.
- Group 2: People you don't know, stranger-on-stranger (criminals).
- Group 3: Enemy soldiers.

Group 1: Friends and Relatives. "To crime report or not to crime report! That is the question." You may end up fighting a friend or relative over, who knows what! Conflicts arise between angry friends, relatives, co-workers, business partners, spouses, often due to sex, money, or personal disputes. When someone you know attacks you, an official crime technically begins, and then they (or you) can become categorized as a victim, or a suspect, or a criminal.

When I began training in martial arts under Ed Parker's Kenpo Karate in the early 1970s, Parker warned that when you fight a friend or relative, you tend to hold back to avoid seriously hurting them. This restraint usually makes the fight last longer. No real finish. I have found this to be true in my experiences and police investigation.

But consider this: if you gouge your buddy's eye out, bite off his ear, crush your uncle's windpipe, or shatter your father-in-law's knee, what happens next? Likely, you'll end up in jail, facing lawsuits, or even dealing with revenge later. Confinement, financial trouble, and legal battles often follow, even if your actions were justified at first.

The *Drunk Uncle Conundrum* asks, "Do you report the crime? Or, does the other person? Or do you both just apply ice to the bruises and move on?" The dynamics of family and friend disputes dictate the response. This is why much violence within personal circles goes unreported, unless there's serious bodily harm. If an ambulance is called, hospitals have reporting requirements that bring the police into the situation, like it or not.

While much violence among family and friends goes unreported, a significant portion does get documented. Domestic violence is a major issue worldwide. The U.S. Department of Justice provides annual statistics, stating:

- "About 29% of women and 10% of men in the U.S. have experienced rape, physical violence, or stalking by a partner of some sort, affecting their daily functioning. Nearly 15% of U.S. women (14.8%) and 4% of men have suffered injuries due to intimate partner violence, including rape, physical violence, and/or stalking."
- Consider U.S. sexual assault statistics: "Crime prevention experts have long emphasized that one of the biggest risks for rape, assault, and murder is who you invite into your home, or whose home you enter. This claim is backed by data showing that 76% of female murder victims knew their killer. In nearly 70% of violent crimes against women, the perpetrator was a relative, friend, or acquaintance. In recent years, females have surpassed males in violent criminal victimization rates."

The majority of violent crimes occur between people who know each other. According to the Bureau of Justice Statistics: Strangers committed about 1.8 million nonfatal violent crimes, approximately 38% of all nonfatal violent victimizations. While these percentages fluctuate yearly, the fundamental reality remains, most violent crimes involve individuals who are familiar with one another. Global estimates from the United Nations Office on Drugs and Crime further highlight the prevalence of intimate partner and family-related homicides.

> "Women and girls in all regions are affected by killings. While Asia is the region with the largest absolute number of killings by far, Africa is the region with the highest level of violence relative to the size of its female population. The Americas (all of the western hemisphere) is next, followed by Europe and Oceania."

Prison Legal News reports the headline: "Most murders are committed by friends and family." *Psychology Today* reports, "Therefore, the statistics appear to indicate that 'the streets' are likely far safer places for women and girls than their home-domestic environments."

Study after study, year after year finds this all true. I consider working homicides as the "World Series of Police Work." I have worked murders for many years and usually the common murder is frequently solvable because it takes some sort of close friend, business, relative, money, sex, serious motive-anger to kill a person. This close motive can usually be researched and uncovered.

The great work of EMTs and emergency room crews can save a life, thwarting a murder, turning murder category into an attempted murder or aggravated assault. Then the victim is alive to tell the tale.

Still, we cannot view the problem of violence by just examining murders and attempted murders alone. Physical conflict is far more complicated and traverses all levels of crime.

Group 2: Stranger on stranger. Someone you don't know. Stranger-on-stranger crime is another kind of beast to solve. I've learned from experience and in many "Assault and Violent Death" schools I've attended, that stranger crime can be very scary to the public. The USA Office of Justice reports, "Stranger-on-stranger crime is often considered more frightening than non-stranger crime because it creates a perception of randomness and lack of control, as the victim has no prior relationship with the perpetrator, making it seem like anyone could be a potential target at any time, regardless of their actions or precautions; this fear is amplified by the unknown nature of the assailant and the inability to predict or prevent such an attack."

In police work, we are often expected to "fight" but not inflict serious harm, unless the situation escalates beyond control. Finding that balance is difficult. In law enforcement, the general consensus is that most arrested suspects comply peacefully. A smaller segment resists to some extent. An even smaller group fights. And the rarest, most dangerous minority attempts to kill you. My experiences align with these proportions. Understanding who we truly fight and under what circumstances is essential for refining self-defense training.

Group 3: Enemy soldiers: Well, we kind of know who those are. Generically, we know the "likes" of the enemy anyway, but as individuals, they are strangers to us. We meet them years later after the war and we can become fast friends. Oh, the irony.

But temporary enemies are often just that, temporary. In war, throughout human history, we have typically tried to kill our enemies from as far away as possible, making it as impersonal as possible. Yet, that distance often collapses, bringing combat nose-to-nose, facial expression-to-expression, blood spattering, coldly personal. Then years later...

Decades later, politics and war passed, oh, the irony... US and NVA officers meet.

High-rate problem groups. In the *Biology of Violence* book, criminal and prison psychologist Dr. James Gilligan explores the nature of violent behavior, "The only two innate biological variables that do appear in violent behavior are *youth* and *maleness*. These patterns are universal across cultures, historical epochs, and social circumstances."

There are many reasons, political, psychological and financial why young men are the target of military recruitment. Malleability. This is not to say that only young males are prone to violence, but it is a crucial observation.

In crime, an individual's age, (as well as size, shape, and conditioning) can be deceiving. In three decades of police work in patrol and investigations, I remained in line operations, deliberately avoiding promotions. In my military police days in patrol and investigations, suspects were mostly young soldiers, yes, but in a unique demographic environment.

Years later in Texas, with years more experience and training, no. Adrenaline-fueled violence is alive and well regardless of age. In my military police days in patrol and investigations, suspects were mostly young soldiers, yes, but a unique, demographic phenomenon. Years later in Texas, with years more of experience and training, no. Adrenaline-fueled violence is alive and well regardless of the age.

Prepare for both friend and foe, recognizing the legal and moral consequences, and maintaining situational awareness. This should be foundational elements of any self-defense program.

WHO Question 3: Who are you legally? Legal Perspectives on Self-Defense: Using force, striking them with hands, feet, a stick, stabbing them, or even shooting depends heavily on "the who's-who"…who is in the situation.

Every country has its own legal framework governing self-defense, crime, and law enforcement. The military, too, has its own rules of engagement. After taking what you received as important action-violence, will you be arrested? Will you be sued? Your past, physical appearance, condition, social media presence, and, most importantly, the situational, circumstances of the event will all shape how the law defines you.

For simplicity, I like to use the Hulk Hogan vs. Pee-Wee Herman analogy. Stay with me on this. Imagine an enraged, unarmed Hulk Hogan attacking Peewee Herman, and Peewee responds by shooting him. Law enforcement, prosecutors, judges, jurors, and the public would likely be sympathetic toward the smaller, weaker Peewee. Now, flip the situation, if Hulk Hogan were to beat up, strike, stab, or, especially, shoot an unarmed, angry Peewee Herman in "self-defense," he would not receive the same sympathy. The public response would likely be: "Come on, Hulk! You didn't need to shoot little Peewee!"

Peewee versus Hulk. A simplistic look at self-defense law.

Yes, this is an *extreme* example, painfully simple example, but it illustrates a core truth: who is genuinely "in fear for their life"? Now, extend this idea across different scenarios:

> Old vs. young, or...
> Infirm vs. able-bodied, or...
> Large man vs. small woman.
> Situation upon situation. Etc. Etc.

These factors influence how self-defense cases are judged, and each case is highly situational. Legal Considerations in Self-Defense Training. Serious gun owners frequently worry about legal issues regarding the use of force, and rightfully so. However, this same concern should extend to hand-to-hand, stick, and knife defense training. On this subject my friend Karl Rehn, a very smart Texas-based, firearms instructor and owner of *KR Training*, makes an important observation:

> "One of the flaws in how unarmed combatives are presented (and how people perceive them) is that most demonstrations in magazines and films involve young, fit males fighting other young, fit males.
>
> Martial arts enthusiasts may believe in the likelihood of winning an unarmed fight, but that's not true for all gun carriers, many of whom are older, weaker, or simply lack training and confidence in their skills."

Extenuating circumstances. Size. age and gender matter in self-defense cases (and all crimes). They create the fully acceptable term "extenuating circumstances," and defined as "factors or situations that can lessen the severity of a judgment or decision, provide context and make actions appear less blameworthy or serious. They are also known as mitigating factors."

Smaller individuals, especially smaller women, may resort to more violent actions against a male attacker, especially a larger male. Fortunately, the law often grants them more leeway in self-defense cases. Ultimately, legal outcomes are determined on a case-by-case basis. But one thing is clear: the way you are perceived with your size, age, strength, mentality and circumstances can greatly influence how the law judges your actions.

WHO Question 4: Who is Trained to fight? So, the fight, or ambush crime begins. Are they trained to do so? After an intense, composed study I amassed evidence, from many diverse sources, through several A.I.s, that roughly 2.5% to 4.5% of the world's population of 80 billion people study martial arts (combat-oriented training only. Not like Tai-Chi.)

About 1 person out of every 40 worldwide has trained-trains in a combat-oriented martial art. And only a tiny subset of that group (perhaps 0.25–1.3% of the world population) trains intensely enough to be genuinely "fight-capable." This info I uncovered could be dissected into its own book. It's vague and short here in this book, but forms a working idea. There are other factors like combat war vets with no martial arts training.

In summary you are most likely to fight an untrained, or partially trained person. This of course is geographically centric. Trained people are "guilty" of fighting other trained people in their art or systems, Boxers fight boxers. Judo fights judo, etc. and expect mirror image attacks and defenses. An untrained person is unpredictable…chaotic. Much more on this later within other specific questions.

WHO Question 5: Who is around you when you argue or fight? We took a look at who you are. We took a look at who you might be fighting. Then we considered the legality of your confrontation. The next question is, who is or isn't around you in an argument, fight, crime or war. Who might be around you in the before, during and after with a knucklehead bully, a mugger, a kidnapper, or attempted murderer, or active shooter, or enemy soldier, anyone attacking you or innocent others?

- *Around you: Your friends.* Will your friends help you? Will they be shocked and freeze? Will they call the police? Will they jump on the back of an attacker and stop them? What is your best guess about each of your friends? It has been my experience from working cases that what they will do might surprise you in both good and bad ways.
- *Around you: His friends.* There are all kinds of studies concerning this subject. One "rule" passed around for years is that 40% of the time you will have to fight two or more people. I cannot find the origin of that suggestion, but you will hear it. Regardless, you may be fighting two or more people.
- *Around you: Strangers.* Are there strangers around? You probably can't count on them to help you or him. But if you are alone and there are people somewhat nearby, try to position yourself to at least be seen or heard. Maybe shout? Keep in mind your attacker might stalk/position you into a spot NOT seen by witnesses. Remember the old saying, "the criminal gets to choose the time and place of his attack. You don't."

We see many videos on the internet now where strangers are caught standing around and watching a nearby attack. They are quickly called cowards for their inactivity. But this might not be fair. There is something called the *Incredulity Response*. If someone witnesses something highly unusual to them, they are unable to believe it because it is very surprising or shocking. They become frozen.

Having mentioned that, there is some positive news. Recently, a group of passengers on American Airlines jumped and duct-taped a crazy passenger trying to get into the cockpit. This reminds us of the heroes on Flight 93 on 9/11, who took control of the plane from the terrorists.

Furthermore, in his history book *Just 2 Seconds* by Gavin De Becker about the history of assassinations, Gavin reports that a number of assassins in world history were quickly jumped by strangers nearby the attack. So, strangers have done well at times to assist protectors. I can understand why the psychology behind each rescue may be different but jump they did.

- *Around you: Fight breaker-uppers and the opposite extreme, fight encouragers.* Strangers or not, yes, there might be those around you who are the peacemakers and try to break up a fight or aid you versus a criminal. Fight encouragers. How many in the vicinity will become excited that you are going to be in a fight, and yell encouragement?
- *Around you: Cooperative witnesses and the opposite, uncooperative witnesses.* Some may stick around and be witnesses for you and later the prosecution. Uncooperative witnesses. Some may refuse to recall what they saw,

wanting to remain out of "trouble," and out of "it." Or incorrectly work against you.

- *Around you: If police and-or military? Teammates.* The Military University Press states, "In the military, teams are trained and expected to prioritize protecting each other during combat, relying heavily on teamwork and loyalty to ensure the survival of their unit members; this is considered a fundamental principle in combat operations, where looking out for your fellow soldiers is crucial to mission success."

National Tactical Officer Association states: "a key principle of SWAT team operations is that team members actively protect each other during missions, relying on coordinated movements, clear communication, and designated roles to ensure everyone's safety in high-risk situations."

While the above teams train regularly to operate in unison, regular police officers do not receive the same repetitive, team-based survival training. They are still expected to cover each other. "Overwatch" of each other is a standard expectation. In all three cases, military, SWAT, and police know the serious consequences of abandoning their partners and the team. (What about sports teammates? They often go to battle for each other, both on and off the field. Absolutely. I certainly wouldn't want to provoke a single teammate among a group of

Australian footy-rugby players out for drinks after a game. No, sir.)
- *Not around: Cameras, surveillance, cellphones etc.* The growing unseen forces that are recording and watching. Film is an incomplete, but important witness. Monitored security services often call the authorities. I have recently seen a cellphone film taken from a distant citizen of a police shooting. It was amazing coverage of a stung-out event. The citizen put it up on the web.

WHO Question 6: Who Will Judge Your Actions?
You've taken some action. Who will emerge from your country's judicial system alive and free? Un-arrested and un-sued? Healthy? Perhaps not so wealthy, after attorney fees, but maybe a little wiser?

So, who will judge your actions, both officially and unofficially? Initially, a police officer will determine whether your actions are justified in the field. Citizens will grumble or cheer. Then, a prosecutor will review the case at their desk. Next, a judge will weigh the matter both in chambers and on the bench. Then a jury will deliberate in their box and decision room. Here are key filters to consider:

Who Judges? Filter 1 The public. The Reasonable and prudent person? "Attorney General Google" defines:
"In the laws of most civilized lands, a reasonable and prudent person is a hypothetical person who acts with the level of judgment, care, and

intelligence that society expects of its members."
"Reasonable and prudent? The law generally means an adult of average age and experience who suffers from no physical or mental disability."

I mentioned in this opening both unofficially and officially. Unofficially, people, the public, witnesses can be troublesome, emotional, and biased. Or not. But there is substantial evidence that eyewitness recollections can often be inaccurate. By now, I hope you've heard of or read about the numerous studies highlighting how forgetful, easily influenced, and unpredictably prejudiced people can be.

After all, a trial is a show, a performance, much like a play. Actors and their acting styles count. And we must consider how unofficial public opinion, shaped by the news and shifting cultural trends, weighs in.

"Public opinion can sometimes conflict with legal principles, forming a 'court of public opinion' where judgments are made without all the facts or adherence to legal standards. This creates challenges for criminal defense lawyers, as pre-trial publicity and preconceived notions can cloud judgment and potentially influence jury decisions."

- Attorney Zack McCready

Public sentiment can turn against you, creating waves of influence that indirectly disrupt your legal defense. If you're tried in a liberal jurisdiction or even in certain countries, you can expect significant challenges. Some of this biased public will inevitably end up on your jury. Good luck.

Who Judges? Filter 2: The Police. Your actions will be reviewed by the police. Ideally, law enforcement officers are professional, honest, unemotional, unbiased, reasonable, and prudent. Their job is to collect and present information about your actions to their supervisors and prosecutors. Consciously or not, they-we apply the classic five Ws and H to assess the situation. However, LGL Law Firm of Florida raises a critical question: "How well do they actually do this?" They echo a well-known mantra: "Law enforcement rarely gives self-defense due consideration during their investigations, and so it is often up to the client and the defense attorney to convince a prosecutor, a judge, or a jury that the act was reasonable self-defense."

Having served for decades as both a patrol officer and a detective, I always wince when I hear the common advice that citizens should remain completely silent on first-police-contact except for saying, "I want a lawyer."

Folks, we the police, need to know at least something about the incident. You should at least state something on the first call or scene like,

"That you did not start the fight."

"That you were defending yourself or others."

"That you are the innocent party."

Etc...

These key, initial details make it into the crime report, which filters all the way to the top. Say something, then politely request an attorney. I was heartened to see that the renowned legal expert and law enforcement veteran Massad Ayoob recently made a short film advocating the same approach. You may still be arrested, but at least your claim of innocence will be "on the first record." More on this later.

Who Judges? Filter 3: The Prosecution. Your actions will next be reviewed by prosecutors in their offices. Hopefully they will be professional, honest, unemotional, unbiased, reasonable and prudent. They prepare the information about your actions for their supervisors, grand juries (in many U.S. states and countries) plea bargains, and for courtroom presentations. They start worrying about appeals.

Who Judges!? Filter 4: The Judges. In the U.S., there are city, county, state and federal judges. Your actions will be reviewed by the court system, either by judges in their offices or out on the bench. There are both criminal and civil (non-criminal) courts.

Key questions arise: Will there be an arrest warrant? A search warrant? A grand jury indictment? Will a civil lawsuit eventually be filed against you in civil court? Judges at the "ground level" approve these legal proceedings, including warrants and indictments.

Even in cases where plea bargains have been reached, which bypass a full trial, judges still review and must approve the agreements.

A well-known 2024 example of judicial oversight involved President Biden's son, Hunter Biden. A plea bargain was proposed that was so insanely lenient it was nearly equivalent to a full pardon or amnesty for ALL he has done. However, the presiding judge reviewed the agreement and ultimately rejected it.

Ideally, judges act professionally, honestly, and without bias, remaining reasonable and prudent in their decisions. They prepare the legal framework for potential acquittals, convictions, and appeals to higher courts.

Civil lawsuits can also arise from criminal cases. For example, in the famous O.J. Simpson case, he was acquitted in criminal court but later found financially liable in a civil lawsuit. The standards for evidence and testimony in civil courts are different, often less strict, allowing plaintiffs to succeed in cases where criminal convictions were not secured.

Who Judges! Filter 5: The Dumbest Juror. Like the citizenry we started this segment with, we hope juries will be honest, unemotional, unbiased, and most importantly "reasonable and prudent." But here's the bad news, many people are simply dense, uneducated, overly emotional, and, sometimes, well... flat-out stupid. And these are the people who end up on your jury.

In a world of declining education, disappearing civics training, combined with low intellectual and emotional intelligence, juries can be a complete gamble. As education levels drop, so too does the general understanding of law and justice. (I should add that I frequently testified in state courts on criminal cases I brought forward. Many times, I could plainly see jurors falling asleep, even though it was obvious to the judge and everyone in attendance. I took it upon myself to cough or fake-sneeze into the microphone to wake them up. Everyone knew what I was doing, but what could they say about a cough or sneeze?)

Who Judges? Filter 6: The Totality of Circumstances. The next level of reasonable, prudent people and judges must consider is the totality of circumstances of your violent action. This standard should exist in all levels, but legal experts say it often plays a bigger part in the appeals process. Attorney General Google reports:

> "The totality of circumstances is a method of analysis or test that considers all relevant factors and circumstances of a situation, rather than using a strict rule. It's more flexible than a bright-line rule and requires courts to consider the whole picture, not each fact in isolation."

For example, consider this scenario: "You shot an unarmed man!" But did you strike, stab, or shoot the first unarmed man because you were being attacked by 3 or more other unarmed men? The situation of an armed individual defending against multiple unarmed attackers is vastly different, and arguably more justifiable, than an armed person simply shooting another single unarmed individual. And recall the previous Peewee Herman versus Hulk Hogan example.

You may have heard that civil and criminal lawyers often "judge-shop," seeking out courts known for being particularly tough or lenient. While the legal system discourages this practice, it happens every day around the world. Additionally, sadly, studies have shown that judges can be influenced by external, unrelated factors, what experts call "noise."

For instance, research has found that U.S. judges tend to hand down harsher sentences on Mondays after their local football team loses. Similarly, they may be overly lenient on a third case after realizing they were too harsh in the previous two cases. They innocently, overtly or covertly respond to human events.

Who Judges? Filter 7: The Afterlife? I will not be discussing judgement in the afterlife here. There are too many versions of this, and I am keeping away from theology and remaining "down to earth." Rest assured many people, myself included, think this judgement is the most important of all. If everyone thought that God was watching them, perhaps they would behave better?

Who Judges? Summary: As you consider the consequences of self-defense situations and survival, keep these critical factors in mind. Discuss them in training groups using the full range of investigative questions, the who, what, where, when, how, and why.

Remember, your fear of attack, of life, must be logical, reasonable, and explainable, especially in lethal force situations. When you strike, stab, or shoot, you must also stop at a reasonable point. Case law on this is clear: excessive force and dishonesty about your fear or the circumstances can severely undermine an otherwise legitimate self-defense claim.

Even if you are 100% innocent, you may face arrest, bail, and the need for legal representation. In the USA, as of 2024, criminal defense attorneys typically charge by the hour, with rates ranging from $100 to $500 per hour, depending on their experience, reputation, and location. Some may also charge a flat fee. The financial burden can reach tens of thousands of dollars, along with the emotional toll of anger, frustration, and anxiety, even if you are ultimately exonerated. It's a rollercoaster ride minus the seatbelts.

WHO Question 6: Who (or what) is suspicious?
This question gets a lot of people in trouble with the subject of profiling. Racial profiling. Religious profiling. Etc. But all kinds of people make heuristic, snap judgements about ALL kinds of other people, all the time. Natural observational concerns are:
1. Who looks like they don't belong?
2. What are they doing?
3. Where are they doing it?
4. When is this happening?
5. How are they doing it?
6. Why are they there?
7. And more of the Ws and H!
    - Who belongs?
    - What are they doing?
    - Where are they going and looking?
    - When do they appear and disappear?
    - How are they dressed, and
    - How do they act?

They don't belong? One of my favorite stories on the subject of profiling comes from my old friend Mike Gillette, an innocent yet insightful experience. Mike, a former police chief and Army Airborne veteran, was hired to evaluate Disney World's security measures after 9/11. During his inspections, he toured the parks nooks and crannies in civilian clothes, looking for weaknesses. As he walked around, people kept approaching him for directions.

Mike is a very cordial, friendly, and charming guy, so after a while, he finally asked one father, "Excuse me, but why are you asking me?"

The father, who was with his family, replied,

"Well, you must work here. You're the only person walking around not smiling or having fun."

Mike was behaving abnormally compared to the other park-goers, not acting like an excited tourist looking for fun. He was simply working. His seriousness, along with his focus on areas that others overlooked, made him stand out.

On the negative side, it's similar to how a shoplifter behaves, looking around too much for witnesses, rather than actually shopping. Mike's actions were indeed out of the ordinary. We, as observers of the abnormal, tend to notice when something unusual stands out.

However, in Mike's case, he did belong, under unique circumstances. Here is a list of suspicious crime prevention tips. My old police department provided this list of suspicious people and circumstances. It's worth reviewing, and this information is also shared on many webpages. Before starting, consider that these tips are also relevant for military security concerns.

- Strange vehicles parked in your area?
- Individuals sitting in a parked car for an extended period of time?
- Person sleeping inside a vehicle?
- Vehicles parked illegally, especially when the driver is inside, idle?
- Vehicles stopped in the roadway or positioned at awkward angles, (possible traffic accident?)

- A vehicle with an open door or trunk with no one around?
- A vehicle with damage such as a broken window or a punched-damaged ignition. A clean vehicle with dirty or damaged plates.
- Stolen vehicles and stolen plates are frequently used in crimes.
- Homeless people or travelers loitering in your area? Their desperation is often your desperation.
- People loitering around your location with no apparent or valid reason.
- People hanging out by businesses after hours.
- Unusual runners. Someone running and dressed "inappropriately" for the exercise, running away, particularly if something valuable is being carried, they may be leaving the scene of a crime. Someone running from a home or business under unusual circumstances?
- Someone with strange mental or physical symptoms, they may be injured or under the influence of drugs or alcohol.
- Anyone forcibly entering a car or home.
- Anybody moving property may be considered suspicious in some circumstances. Perhaps if it is late at night or if the item is not wrapped, it is possible that the article has been stolen.
- Somebody going from house to house, such as a salesperson. This is particularly suspicious if, once a few homes have been visited, one or more of the people go into a back or side yard. It is even more suspicious if another remains in

the front when this occurs. It may be that they are looking for a house to burglarize, or there could be a burglary or home invasion in progress.

- See people in private communities who do not live there. These communities usually require the presence of a resident or tenant to escort people not living there. Without this escort they could be considered trespassing.
- Individuals who don't fit into the surrounding environment because they are wearing improper attire for the location or season. Concealing weapons or just crazy?
- People "piggybacking" behind people entering a gated or authorized only area. Walking or driving? Is it a maid? A worker? Or...
- Someone carrying a weapon in an inappropriate setting. Like walking through a school, mall or church carrying a rifle with an angry demeanor.
- Individuals drawing pictures or taking notes of an area not normally of interest to tourists, or showing interest in or photographing security cameras, guard locations, or studying security procedures.
- People who may try to have a "cover story" or appear 'normal' in their behavior such as portraying a student, shopper or tourist.
- Many comings-and-goings from a particular house. This may not be suspicious, unless it occurs on a daily or very regular basis, especially during late or unusual hours when it could signify vice or drug related activities.

- Multiple sightings of the same suspicious person, vehicle, or activity. Unusual requests for information, particularly about security or procedures for at-risk buildings.
- Testing local residents by breaching restricted areas to determine if anyone will react.
- Tampering with electrical, water, gas, or sewer systems. Criminals may conduct training, surveillance and "dry runs" prior to an act.
- Criminals may conduct surveillance to determine a target's suitability for attack by assessing the capabilities of local residents in discerning potential weaknesses.
- Unusual rentals, purchases, deliveries, or thefts, particularly of poisonous or flammable chemicals, explosives, weapons or vehicles (including planes or boats).
- Anyone ringing a doorbell or knocking on a door without a reasonable explanation for doing so.
- Unusual or extended interest in public buildings, utilities, large public gatherings.

Keep in mind so many of these 'pre-incidents' or suspicious activities are not crimes, yet? These behaviors may not be illegal in and among themselves, but suspicious in varying degrees. What steps suspicious people do to confront, ambush and attack you is a HOW question to be covered later in that chapter.

WHO Question 8: Who will take avenge, revenge, vengeance? Do the words avenge, revenge, and even vengeance, have distinct differences between them? Experts at Grammerist.com thinks so.

- Avenge: To avenge someone means to take action in response to a wrong done to them out of a desire for justice, while revenge is motivated by anger or malice.
- Revenge: To revenge is most typically used as a noun (even though it is also used as a verb sometimes), while avenge is a verb. Revenge, reprisal, retribution, vengeance. Revenge wants a punishment, or injury inflicted in return for one received. Revenge is the carrying out of a bitter desire to injure another for a wrong done to oneself or to those related.
- Vengeance: It's a noun, and it means the punishment delivered or the retribution you get for an offense or injury."
- Avenge, revenge, and vengeance are often confused due to their similar meanings. Both avengers and revengers seek to right a wrong, but their motivations differ, avengers act in pursuit of justice, while revengers seek personal satisfaction. Vengeance, on the other hand, is the result of carrying out either or both of these actions."

History, especially in times of war, is filled with stories of avenge, revenge, and vengeance. But do we need to turn this book into a history textbook? No. But, the same themes run through crime history as well. Should this become a book on global crime? Again, no.

I have investigated numerous cases related to these three concepts, sometimes actions taken decades after the original act of violence. And such are not always taken by the injured party alone; they may also be carried out by friends, family members, or sympathetic groups.

Mistreatment, and certainly violence, always brings both drama and trauma. Those caught in such cycles must remain wary, watching their backs for a long time... perhaps even forever.

> "Before you embark on a journey of revenge, dig two graves." – Confucius

> "That old law about 'an eye for an eye' leaves everybody blind. The time is always right to do the right thing." - Martin Luther King, Jr.

WHO Question 9: Who will teach you to think, talk, fight, stab, shoot? When selecting a martial system and instructor, one must be cautious. Many martial arts are marketed as effective for self-defense, but this is misleading. Some are primarily sports oriented. Others focus on artistic expression with abstract benefits. Combatives and survival-based training should not be sport-like or artistic, yet some programs suffer from excessive and legally questionable tactics. For example, a college professor friend once told me about his experience in an Israeli martial arts class. The instructor required students to perform a self-defense sequence

 that involved gouging out an attacker's eye, slamming them onto the pavement, and repeatedly kicking their head. He noted that this level of hyper-violence was consistent throughout the course training, with no discussion of the legal ramifications or appropriate use of force, just pure overkill.

I recall another case involving an overzealous karate school. A homeless panhandler became a little too aggressive while begging for money from a male supermarket shopper, who happened to be a student at the school. The student brutally attacked the panhandler in the parking lot and was sentenced to two years in prison. He was eventually granted early parole, but his instructor's response was, "That'll teach those beggars to leave people alone." Seriously? Your student just spent a year in prison! Perhaps it's time to reconsider what, how, and why you teach your students.

Aspiring martial artists often feel limited by the schools available within a convenient five-mile radius. However, to receive the best training, you may need to travel. Take the time to research courses and instructors in hand-to-hand combat, stick fighting, knife defense, and firearms. The questions outlined in this book will help you make an informed decision. Who then in these lists, inspires you?

WHO Question 10: Who inspires you? Everyone needs inspiration. Mentoring. Though I'm in my 70s, I still joke around and say, "When I grow up, I want to be like _____." (I am after all. much grown up.)

Actor Matthew McConaughey once said during an awards ceremony, after crediting God and his family:

> "You see, every day, every week, every month, and every year of my life, my hero is always ten years away. I'm never going to be my hero. I'm not going to obtain that, and that's fine with me because it keeps me chasing someone."

Dan Inosanto, the martial arts world's equivalent to Elvis Presley, once shared a story in a seminar back in the 1980s. Forgive me for forgetting the exact football names. It's been a while. He recalled his time as a high school running back when he frequently ran straight into big defensive tacklers. His coach pulled him aside and asked, "Dan, why do you run into these guys like that?"

Dan replied, "Well, Coach, I want to be like my hero, Joe _____."

The coach shook his head and said, "Dan! Joe is 6'4." And you're 5'5." Pick another hero!"

Dan did just that and he chose a smaller running back as his inspiration and went on to break California state high school running back records. Same guy, better hero choice. Mimic the possible.

I love stories like that. Who inspires you? Maybe like me, it's Jesus, Martin Luther King Jr., Ernest Hemingway, or a sports star. You can have many inspirations, one for each field, topic, or endeavor. But always remember to look up, not down.

How can I perform better? Inspirational mimicry in performance. Mimicry in performance is a classic and powerful tool. Inspirational people move us, mostly mentally. Mimicry, on the other hand, is about imitating them, mostly physically.

I have recalled a story from Dan Inosanto about his high school football coach. The coach told him to mimic a hero who matched his size and shape. He did. Since we're on the topic, here's another Dan story: His coach also made the team run up and down steep hills during practice. When running downhill, they were instructed to run as fast as possible. Since the incline forced them to move faster than they would on flat ground, it was like running right on the edge of falling. Inosanto described it as "a bit like flying." His coach told them to "remember that feeling of that speed. Mimic it."

Recently, I watched NFL highlights and saw the incredible Saquon Barkley sprint about 50 yards for a touchdown. He ran toward the camera, his facial expression inside the helmet was beyond serious. His machine-like, chainsaw-leg speed was blistering. It was inspiring, powerful and I thought, at the very least, I could mimic that intensity. I tried it during my routine runs. While I'll never be like Barkley, however I did run faster. He inspired me. I mimicked him.

Growing up in the New York City area, I had the opportunity to play baseball and football endlessly through countless organized leagues. I played baseball until I was 18, competing alongside men in their 20s and 30s on money-sponsored teams. (I left NYC, just turning 18 years old on a motorcycle.)

During those years, I, like my peers, was under the influence of numerous coaches. But back then, few had any real expertise in performance coaching or a "teaching IQ." Even in high school, where baseball, football, and wrestling were huge, coaching mostly consisted of some classic routines, tips and occasional mistake corrections, nothing like the innovative additions in training we see today.

They'd say, "Playing is learning," much like the martial arts phrase, "Learn to fight by fighting." But that method only takes you as far as your natural athleticism allows. Think of the prematurely athletic teen who was always the high school's star quarterback, yet 99.999% of them end up working regular jobs after high school. Without skills coaching, they maxed out their potential early.

So how did any of us improve? Most coaches back then didn't have the tools to push us forward. It seemed like mimicry was a primary teaching tool. "See how so-and-so does it? He's successful. Copy him."

One of my personal goals was to play third base, the "hot corner." The position demanded quick reflexes, a

strong arm, and the ability to handle fast ground balls and line drives.

To improve at third base, I mimicked Yankees' third baseman Clete Boyer. One day, I saw Boyer dive full-out for a fast foul ball. It was foul, why not let it go? I realized that in his mind, at the crack of the bat, every ball was his to catch. He didn't stop to analyze whether it was fair or foul. His instinct was simple: Ball. Catch. That mindset inspired me, an epiphany. No coach had ever explained it that way. So, I mimicked Clete Boyer. But without today's structured coaching methods, I could only progress so far. Epiphanies, those rare moments of inspiration, are too few in life. Too hard to orchestrate and regiment. A bit magical.

Teaching IQ? Teaching technology? Consider Willie Stargell. I followed his career with the Pittsburgh Pirates and came across a bio book detailing his late-career resurgence. Like many aging athletes, offseason, decades ago, Stargell paid his own way to one of the very first, official baseball training camps, an uncommon practice at the time. his performance began to decline. The Pirates considered moving him to first base, or even benching him. But that

At the camp, coaches and Stargell identified every possible in-game scenario, every event, every situation he might face. Catching the ball a certain way, running from first to second, first to third, they cataloged about 30–40 specific actions. That winter, Stargell drilled each one 100 times a day under their watchful eyes. The result? A career resurgence. He extended his playing years and continued attending that camp every offseason until age finally caught up with him.

Wow, another epiphany for me. Tools. Steps. Isolated skill exercises. It changed how I thought about coaching and skill development. Today, this methodology has spread widely, even at the grassroots level. When I coached my kids' baseball teams, I made it a point to pull players aside and work on isolated specifics.

Over the past few decades, we've seen these structured methods improve every sport. Records are broken regularly. The talented refine their abilities. The mediocre get better. Even those who once struggled still find improvement. Entire buildings are now dedicated to youth sports training, much like the once-primitive school Stargell attended.

"THOUGH PUNCHING TRAINING IS MUCH THE SAME, MIKE TYSON SHOULD NOT FIGHT LIKE MOHAMMED ALI, AND VICE VERSA..."

MIMICRY IN PERFORMANCE

And what is this training, if not another form of mimicry? Mimic the training methods of great coaches. Are you too tall and lanky to be a baseball catcher, a gymnast, as I was? There's a reason successful they have a particular build. Wrestlers too. We all have different physical attributes. Basic boxing principles remain the same, but Mike Tyson should not fight like Muhammad Ali, and vice versa. Dan Inosanto should not run the football like a giant fullback. Isolate skills. Build them. Oh, and allow me to remind you of Mirror Neurons...

> "Mirror neurons represent a distinctive class of neurons that discharge both when an individual executes a motor act and when he observes another individual performing the same or a similar motor act." - PositivePyschology.com

Looking for inspiration, ask yourself the fundamental questions:

- Who are you? Your size, shape, age & strength matters. Who should you mimic? Who are your heroes, and to what end?
- What is your goal, and what methods exist to achieve it? What should you realistically expect?
- When can you train? (Remember, Willie did it in the offseason!)
- Where can you work on this?
- How can you pull this training off?
- Why do you want to do this?

Back to aspects of the WHO question. Who do you inspire? As a father, mother, spouse, grandparent, supervisor, teammate, deacon, pastor, teacher, professional, friend, or public figure, who looks up to you?

I will leave the subject with this. Perhaps some inspiration? As an older child and young teen, I read many Sherlock Holmes stories and detective books by Raymond Chandler, creator of the quintessential P.I. Phillip Marlowe. *In the Art of Murder* book, Chandler wrote:

> "Down these mean streets a man must go who is not himself mean, who is neither tarnished nor afraid. The detective must be a complete man and a common man, and yet an unusual man. He must be, to use a rather weathered phrase, a man of honor by instinct, by inevitability, without thought of it, and certainly, without saying it. He must be the best man in his world and a good enough man for any world."

This iconic paragraph moved me as a kid. Stuck with me. Sketched out my life. A code, a theme for life. A code. Code of the West. What sketched you out? And by the way, your painting isn't done yet. It's still a little wet. It ain't dried until it becomes a tombstone.

"MY HEROES HAVE ALWAYS BEEN HEROES. WHEN LIFE AND TIMES RAN THEM RAGGED, THEY PERSEVERED. THEY DID. I TOO WILL TRY TO PERSEVERE. BECAUSE? MY HEROES...HAVE ALWAYS BEEN HEROES." - HOCK HOCHHEIM

WHO Question 11: Who will stand beside you through your trials and tribulations? Can you predict who in your life will? Spouse? Family? Friends? Parents? Conflict and certainly more so, violence is a drama and a trauma for those involved. And it can be very expensive also.

The WHO Summary:
There are numerous self-evaluation tests available. Find and take a few if you wish, but keep in mind that results vary depending on the methodology used.

As you continue your journey, keep developing and asking more "Who" questions. You'll likely find yourself revisiting this question multiple times, as illustrated in the following "deep dive" section.

Keep exploring both the small and big "Who" questions and their answers.

> **The Who Question Review.**
>
> Who are you within this who category?
> What are you within this who category?
> Where are you within this who category?
> When are you within this who category?
> How are you within this who category?
> Why are you within this who category?

Chapter 4:
The What Question and Confrontations

The "What if?" and What Happens Next" Continuum. A confrontation, a crime or a battle has occurred. You may have started it, or he-she-they started it. Perhaps the most important what question is "what happens to you next?" Big and small.

> What made you go there?
> What made you stay there?
> What will make you become arrested?
> What will cause you to be sued?
> What will happen that you will become hurt?
> What will happen you will be killed?
> Will you get back home safe, or back to your base/headquarters safely and change your socks? (Military people know how important socks are.)

Small "what happens next" moments in confrontations can be subtle like certain looks, specific words, slight shifts in positioning, a feint, hands slipping into pockets, a twist of the torso, or a crouch at the knees. These small signals can tip you off, prompting a decision. Should you, can you leave before trouble escalates? Prepare for physical contact.

Evacuate or escalate? What action, small or big, will you take when you notice a subtle sign of impending trouble? In all areas of life, every step you take, or don't take, has consequences. Whenever possible, consider:

"What will happen next if I act?"
"What will happen next if I don't?"

WHAT Question 1: What is the big picture of safety and survival? Usually missing inside any news story or opinion piece is the big picture statistic, which almost always places things in a better perspective. This news is often fear mongering by way of "out of context" broadcasts. I shall offer you some quick examples to kick off this chapter. (There will be a few more "big picture" studies like this, in subject matter categories to follow.) Grasp my intention to respect the "big picture."

Big Picture Look at vehicle accidents in the U.S.A. There are a lot death and injuries from this. But…
1. Total cars in the USA as of January 2025 were 288.5 million.
2. Trips. A healthy portion of the 288.5 million autos are on the road every day in small and big trips. Let's just pick a rounding number like 2.5 million of the 288.5 million, a conservative number taking daily trips. That equates to a

very conservative number of 800,000,000 trips a year. A nice estimate.
3. 40,990 people died in motor vehicle traffic crashes in 2024. That's about 41,000 deaths in about 800 million vehicle trips.
4. Run the numbers in your country. Total cars, total daily trips, total crashes, total deaths.
5. One might read a news report about 40,990 motor vehicle fatalities a year and freak out, yelling/demanding, "The highways and byways are too dangerous! The Department of Transportation must do something!"

The fatality headline alone sounds like a nightmare, yet it fails to mention the incredible, total amount of cars and safe travel. The big picture is missing. Of course, this doesn't mean we shouldn't be the safest drivers possible. This is the same for air travel. Very safe.

More big pictures - U.S. places of worship and crime.
There are roughly 400,000 churches and approximately 3,700 synagogues in the United States. People meet there all through the week, 52 weeks a year. That's a whole lot of successful meetings at a lot of churches, juxtaposed to the teeny amount of church shootings. Tragic? Yes. That doesn't mean there is no need for church security. It's just a look at safety percentages.

The United States has many schools and universities, including colleges, private schools, and other educational institutions:

1. There are approximately 6000 colleges and universities in the US, including 1,626 public colleges, 1,687 private nonprofit schools, and 985 for-profit schools.
2. Private and Catholic schools make up about 17% of US schools. These schools include 13,871 private (non-Catholic) schools and 5,458 Catholic schools.
3. There are also 1,985 Title IV non-degree-granting institutions. As of 2021, there were 30,160 high schools in the United States. This includes both public and private schools.
4. There were 88,925 elementary schools in 2019-2020, which included both public and private schools.
5. The average number of U.S. school shootings varies significantly by definition and year, with figures ranging from around 100-200+ annually. Though definitions differ on whether to include active shooter events, all gunfire on campus, or just those with casualties. Some jurisdictions send in a school shooting stat for a weekend nighttime, drug deal gone bad. In comparison, that's a lot of schools successfully meeting weekdays for most months a year, juxtaposed to the teeny amount of school shootings.

Tragic as they all are? Yes. This doesn't mean there is no need for school security, no. It's just a look at big picture safety percentages.

More big picture looks: Guns in the USA! As of 2026:
1. The population of the United States is some 346,487,685 people as of December 2025.
2. The United States has between 400 and 500 million civilian-owned firearms.
3. This is a sketchy number, but in 2024, approximately 44,000 people died from firearm-related injuries in the U.S. This is about average, year to year, and a tragic, yet teeny number in comparison to the population and proliferation of firearms.
4. Mixing those numbers, we see that gun crime is quite teeny in comparison to the population and the number of guns. Each event is tragic, yes. Of course. No, that doesn't mean we shouldn't aggressively enforce gun-related criminal laws. It's just a look at safety percentages (I was interviewed on BBC in England a few years back and I dared to say to the anti-gun host, "If you believe all these anti-gun liberals, we should all be dead in the USA by now." I was being only slightly facetious, but of course the liberal announcer did not like that answer. Not at all.)

More big picture looks. Police killings. After the George Floyd miasma in 2020, a "black-men-killed-by-cops" protest wave began. Many protestors were quizzed on film and radio about how many black males they thought were shot and killed by police each year in America. The answers were astoundingly wrong. Some said "about 10,000 a year?" "Thousands a year?" The highest I recall seeing-hearing on the news was one couple saying "14,000 a year?" with ending question marks? The ignorant did, each time I watched and heard them, all end with a "question-mark-guessing" tone. They knew they didn't know how many, yet cities burned. Some people need excuses to burn.

Experts say that the Washington Post did the best ten-year study on police shootings-killings from 2015 to Dec 31, 2024. Let's assemble some information, as best collected. The Post admitted the collection data streams are not very efficient.

- Population through the 10 years, for reference.
  - Population of the US in 2015 totaled 320 million, growing to 340 million in 2024.
  - Population of Blacks in 2015 totaled 42 million, growing to 56 million in 2024.
- 2,486 blacks were shot-killed by police in 10 years.
- 1,717 whites were shot-killed by police in 10 years.
- 4,659 Hispanics were shot-killed by police in 10 years.
- 380 "others" were shot-killed by police in 10 years.
- The total is 9,243 killed within that population realm of 320 to 340 million people over ten years.
- Rounding off the ten years to a 330 million population, that is 0.0028% of the population shot and killed by police. Race-baiters, ignorant rebels, Antifa, the media, etc. will always play with the per-race percentages.

- The Post and other sources such as the Guardian report that it is well accepted that most of these killings involved armed suspects (knives and guns). Exactly how many, they cannot exactly say? Why? (One reason is info collection problems. They do exist.)
- People that were shot and did not die numbers are too elusive to collect.

Each event is tragic, sure. All violence is both a trauma and a drama. But the radical Floyd response was ignorant and political. And made millions of grift for BLM leaders. Look it up. Burning and looting cities is criminal, counter-productive and misguided.

The big-big picture safety summary. Let's not forget the big picture. What's going on in your country? Are you served up the big picture, good, clean news with regularity? Or just the scant, bias, terrible headlines that frustrate, twist and torture our brains? Are you spoon-fed propaganda media inside a political bubble? Fear mongering with bad news out of context for agendas? Run every event through the who, what, where, when, how and how investigation using multiple sources.

WHAT Question 2: What makes you a target? A potential victim? And what exactly is victimology? Victimology is the study of victims before, during, and after incidents of victimization. It examines the patterns, causes, and consequences of being victimized. As a branch of criminology, victimology focuses not only on the crimes themselves but also on the experiences of those harmed by crimes, accidents, or abuse.

One of the most influential frameworks in victimology is the Routine Activity Theory (RAT). It was first introduced by Marcus Felson and Lawrence Cohen, who used it to explain shifts in U.S. crime rates between 1947 and 1974. Today, it is one of the most cited theories in criminology.

Unlike theories that focus on criminal personalities, RAT treats crime as an event. It emphasizes the ecological and situational nature of crime, linking it directly to time, place, and opportunity rather than only to offender motivations. Importantly, RAT often uses the term target instead of victim, since the potential victim may not even be present at the crime scene.

For a crime to occur, three elements must converge in the same place at the same time. This is often illustrated as the RAT Triangle:

- Motivated Offender. A person with the intent and ability to commit a crime.
- Suitable Target. A person, object, or property that is visible, valuable, accessible, and easy to move or exploit.
- Absence of a Capable Guardian. A lack of substantial nearby help, deterrence, such as police, security, neighbors, or surveillance.

RAT provides tools for crime prevention. For example, increasing guardianship (e.g., security cameras, better lighting, active bystanders) or reducing target suitability can significantly lower risks.

Who Is more likely to be victimized? Victimology research identifies several factors that increase a person's likelihood of victimization:

- Age – Young adults, especially those 18–24, are at the highest risk for violent crimes. Older adults, however, are more vulnerable to scams, fraud, and elder abuse.
- Gender –Men. More likely to be victims of violent crime (assault, homicide).Women: More likely to experience domestic violence, sexual assault, and stalking.
- Lifestyle / Routine Activities. – Going out at night, frequenting bars or clubs. Associating with high-risk groups adds exposure to offenders.
- Socioeconomic Factors. – Poverty, unstable housing, and disadvantaged neighborhoods raise risk due to weaker community security and fewer resources. Personal & Situational
- Factors – Substance abuse, disabilities, or mental health issues can make individuals easier targets.
- Repeat Victimization. – Once victimized, individuals are statistically more likely to be targeted again.
- Environmental Conditions – Poor lighting, lack of surveillance, and living or working in high-crime areas (e.g., taxi drivers, convenience store clerks, sex workers) elevate risk.
- Routine Exposure to Strangers. – Daily reliance on public transportation, crowded spaces, or online platforms increase opportunities for offenders.

Personal Vulnerabilities. – Isolation, intoxication, lack of self-defense skills, or dependence on caretakers can make individuals easier prey.

The Role of Appearance and Behavior. Criminals often select targets based on signals of weakness, confidence, and awareness. Body language and appearance can either deter or attract offenders. factors like:

- o Posture & Presence. Standing tall, walking with shoulders back, and maintaining situational awareness signals confidence and reduces vulnerability.
- o A hunched posture or distracted behavior (e.g., headphones, phone use) may suggest an awareness weakness.
- o Eye Contact – Calm acknowledgment of others shows awareness, discouraging predatory targeting.
- o Movement Speed – Moving significantly slower than others can suggest injury or distraction, much like wild predators isolating the weakest animal in a herd.
- o Even appearing athletic can deter offenders.
- o Clothing & Environment Fit. Dressing appropriately for surroundings helps avoid standing out. In military and tactical contexts, specific clothing has historically marked individuals as targets, sometimes fatally.

- Parental Strength as Deterrent. Interviews with convicted pedophiles revealed they often assess the father's presence. If a father appears weak or non-threatening, the child is more likely to be targeted. A strong, protective presence, on the other hand, deters offenders.
- Conclusion. Victimology and Routine Activity Theory together highlight that victimization is rarely random. Instead, it emerges from the convergence of opportunity, vulnerability, and environment. By understanding who is most at risk, and how appearance, behavior, and lifestyle shape that risk, individuals and communities can take practical steps to reduce victimization and strengthen guardianship against crime.

Awareness and education. I have always been fascinated by "Rumspringa." In some Amish communities, teenagers undergo this rite of passage, a time when they are encouraged to explore behaviors that are normally forbidden or restricted. The term Rumspringa is German for "to run around" and typically begins around age 16, though it can last until age 24. The essence of Rumspringa is not just experimentation but discernment and exposure.

While some may view it as an opportunity to indulge in modern vices, its true purpose is to allow young Amish individuals to experience the outside world and make an informed decision about their future. These near-adults, or in some cases full adults, have had little to no exposure to crime and other harsh realities of life. As a result, their understanding of what is normal or abnormal is extremely limited, making them particularly vulnerable.

On their journey, they are more susceptible to deception and crime. Criminals often prey on the inexperienced, exploiting their naivety. (For those interested, The Journal of Rural Social Science published a paper titled Amish Victimization and Off Amish Victimization and Offending: A Subculture's Experiences and Responses to Crime and Justice, which delves into this issue in detail.)

Now, imagine navigating life with almost zero real-world experience. Consider everyday citizens enlisting in the military. Most recruits have no working knowledge of soldiering, battlefields, or war. They shouldn't be thrust into combat any more than a baseball player should be forced into professional rugby. Or a plumber? They must first be exposed to, trained for, and educated in their mission.

Criminals (and enemy soldiers) take time to set you up. While some criminologists claim that criminals decide in just seven seconds whether someone is a victim, this is not always the case. Maybe a quick street mugging may be decided in seconds, but many other crimes and attacks involve meticulous planning, sometimes over days, weeks, months, or even years.

Be aware of patterns. Some crimes require little preparation, like a spontaneous mugging, while others involve extensive surveillance. Certain locations and times make individuals more vulnerable. Stay mindful of your daily, weekly, monthly, and even annual routines, as criminals often exploit you and your predictable behavior. Some tips:

- Armed robbers love paydays, armored truck deliveries, and robbing stores full of Christmas purchase monies.
- Burglars love to know when you leave your house.
- Home invaders love to follow you home from work or expensive restaurants and duck into your garage door before it closes. Safety experts suggest changing those travel routes, schedules, routines when you can. I have other suggestions listed in this book.
- Muggers and rapists love to watch the ebb and flow of pedestrians.
- Always check your local and regional news for crime patterns.
- In crime and war, moderate to severe winter usually slows things down or stops things.
- You can see different crimes need different times of day, weeks. even years.
- Think like a criminal. To solve crimes as a criminal investigator, I had to start at the crime and work backwards, asking these Ws and H questions. One is "what did it take to dream-up, set-up this crime?" For prevention purposes you must start from this "what did it take to do this crime." Then calculate your weaknesses in the future, making yourself safe, crime by crime.

Situational awareness: Perhaps you've grown tired of the rather overused term "situational awareness," being constantly reminded to "stay alert!" But have you ever truly considered what it means to be aware of how you act, how you look, where you are, and who you're with?

The word awareness alone simply means being conscious of your surroundings. As discussed earlier, we are naturally attuned to recognizing the normal, and most people instinctively notice when something seems off.

However, the brain must also be trained to detect subtle nuances. The addition of the word situational refines this concept further, emphasizing a more deliberate and practiced awareness. SpideySense!

What's your Spidey Sense? Captain Google defines it as: "In Marvel Comics, Spider Man's, Spidey Sense is an extrasensory ability that allows him to sense and react to danger before it happens. It's often depicted as a feeling or premonition that something is wrong or about to occur. Spidey-Sense can also guide Spider Man's movements, allowing him to focus on other problems." Oh, to have an innate Spidey Sense!

But we don't. The so-called gift of fear can only take us so far.

Sensemaking. Our minds constantly collect information about public safety, crime, and conflict from countless sources, books, television, news, movies (both fiction and documentary), training, courses, personal experiences, mentors, and formal education. We were not born fearing dark alleyways; we learned to. Over time, we build a mental filing cabinet, a reservoir of knowledge, sometimes reliable, sometimes flawed, that shapes our perception of threats and helps us deduce what to be cautious about.

This process is now called Sensemaking, not just seeing or understanding, but actively analyzing and interpreting situational awareness. It goes far beyond merely being aware; it is about processing information and making informed decisions based on context.

The term Sensemaking has long been used in the business world, but it has now been adopted by U.S. military programs. The military once pursued the now-infamous Jedi Project, a failed attempt to harness psychic abilities for warfare (see The Men Who Stare at Goats or read Annie Jacobson's book Phenomena actually, read all her books). Realizing that training soldiers to be psychics was a dead end, and they shifted their focus to developing sensemakers, individuals trained to logically assess military situations and make sense of every factor and player involved.

- Researchers at Elsevier and Science Direct "Sensemaking is the process of developing an understanding in the face of unexpected information. Military leaders often face uncertain or unfamiliar situations and must be able to quickly make sense of them to take appropriate action. Sensemaking can occur in combat and after combat and can involve different aspects for different ranks. Intuition is often thought of as a single construct, but our 2-year longitudinal study of multiple managers developing opportunities uncovered four distinct types of intuition:

> 1: expert intuition, based on previous experience.
> 2: creative intuition, based on a sense of direction, for a novel solution.
> 3: social intuition, based on a sense of interpersonal relationships.
> 4: temporal intuition, based on a sense of the timing being right to create or capture an opportunity."

The sensemaking art, science, and methodology of understanding complex situations is becoming a part of college curricula and serves as a valuable tool for enhancing situational awareness. It inspires individuals to analyze situations and make informed deductions across various fields, including business. In fact, the marketplace has largely integrated sensemaking into the corporate world. However, it also plays a crucial role in military and law enforcement operations, as well as in civilian survival and self-defense.

One military lesson plan I encountered involved a foot soldier in a platoon cautiously entering a questionable village in Afghanistan. The soldier was instructed to observe and evaluate everything, the commander's conversation with village leaders, bystanders' reactions, and the surrounding environment. This level of heightened awareness is fundamental to being an experienced and alert soldier.

Of course, such sensemaking requires sufficient intellectual and emotional intelligence, traits that unfortunately, not all citizens or soldiers possess. Even without an innate "SpideySense," developing a structured approach to sensemaking can help prevent you from becoming a target by rapidly assessing and understanding unfolding situations.

WHAT Question 4: What is Winning Anyway?
Defining victory: What Are You Really Fighting For? First off, what is your mission? What is your ultimate goal? What are you truly fighting for, and what are you not fighting for? Many training courses demand that you "win!" But what exactly does victory mean?

In the past decades, countless police, military, and civilian instructors have preached the "winning is everything" mantra. While this message may seem straightforward, it can sometimes be misleading or even dangerous. I believe this mindset oversimplifies the complexities of warfare and crime situations. Instead, we must broaden our understanding of what it means to win. We need to focus on situational problem-solving.

As I mentioned earlier, sometimes we need to step back and reassess. Who are you in this moment? Are you a police officer? A soldier? A civilian? What is your specific role, and what is your objective?

- A civilian may define victory as escaping an attempted mugging in a parking lot, unharmed.
- A police officer may define victory as successfully arresting a combative suspect and not be charged with excessive force.
- A soldier may define victory as capturing or eliminating an enemy.

Whether it's evading a mugger, avoiding a confrontation, making an arrest, or engaging in combat, victory means different things to different people, in different situations, at different times. Winning is always situational. It depends on lifestyle, mission, and circumstances. When Is Retreat the Right Move?

Does winning mean always standing your ground against impossible odds? Imagine:

- You're part of a six-man foot patrol in the Mekong Delta during the Vietnam War. Suddenly, an entire battalion of North Vietnamese soldiers surrounds you. Do you fight to the death or try to escape form overwhelming odds
- You're a city police officer patrolling alone when four armed assailants ambush you.

- You're a civilian trapped in a supermarket parking lot, surrounded by cartel gunmen.

In all these scenarios, it's clear that discretion may be the better part of valor. Sometimes, the smartest move is to live now so you can fight another day, when you have a real chance to win. The best warriors know the value of an orderly retreat.

Experienced operatives even develop escape plans in case everything falls apart. Some overseas contractors I know joke... "When in real doubt, head for the airport."

The Wisdom of Retreat. You've probably heard the phrase:

> *The better part of valor is discretion, in the which better part I have saved my life."*
> *- Falstaff, Henry IV, Part One*

Yes, Shakespeare strikes again. This phrase reminds us that caution is often wiser than reckless bravery. Despite all the motivational "never back down" speeches, every warrior, whether fighting alone, in a small unit, or as part of a massive force must recognize when it's time to retreat. And when retreating is necessary, it should be done properly and orderly.

So, what is winning? So, who are you? So, what does winning mean to you? Is it surviving? Escaping? Accomplishing your mission? Regardless of whether you are in law enforcement, the military, or a civilian defending yourself, the reality is this: Situational combat, whether in war or crime, requires flexibility.

Victory is not always about standing your ground at all costs. Winning might be:

- Escape from the opponent (using the "Orderly Retreat" concept. See the next essay).
- Threats, demands, and actions to make the opponent surrender and/or desist and maybe even make him leave.
- Less than lethal injury to the opponent. Injure and/or diminish to a degree that the opponent stops fighting, and/or stops chasing you.
- He leaves. No physical contact. You use threats, demands and actions to make the opponent desist and leave.
- Control arrest, contain, and restrain until help arrives.
- Physical contact. You inflict less-than-lethal injury upon the opponent. Injure and/or diminish to a degree that the opponent stops fighting and won't chase you.
- Lethal methods. We fight criminals and enemy soldiers. Sometimes we kill them.
- Know your mission. Having a cause is a great motivator.
- Study the who, what, where, when, how, and why of your life and make solid, retreat plans.
- Winning and the law don't always align.
- Sometimes, the smartest warriors win by knowing when to retreat.

WHAT Question 5: What then is an orderly retreat?
If trouble arises, the law will scrutinize your actions. First it asks, why were you there? Next question, why did you stay there? And then comes the moral dilemma, can you leave behind your family, friends, comrades, or innocent bystanders? Should you?

Sometimes, leaving simply isn't an option. Your decision is highly situational. Whether facing a fistfight, a crime, an active shooter, an ambush, or even full-scale war, always ask yourself: Should I stay or should I go?

Many so-called safety experts casually advise, "Just turn and run away!" But how far can you run? How fast? For how long? And more importantly, how far, fast, and long can your enemy?

An orderly or tactical retreat must be clearly defined, with precise steps for safely withdrawing, whether walking away or running. Should you stay or retreat?

A retreat has multiple definitions, from the traditional military "retrograde" movement to the more modern concept of a tactical retreat. However, in today's civilian world, choosing to stay and fight can lead to serious physical, legal (both criminal and civil), and financial consequences.

Retreating ultimately means leaving, escaping, or withdrawing, ideally with a well-thought-out plan. In smaller, personal situations, an orderly retreat means exiting a confrontation safely, without provoking pursuit or escalating the situation.

For a classic example, we can learn from Alexander the Great. His army suffered remarkably few casualties while reportedly killing up to 1.2 million enemy soldiers, a staggering difference. How? In his early battles, Alexander's forces suffered heavy losses, often due to chaotic retreats. When his troops reached their perceived "breaking point," they scattered in panic, and most were cut down from behind. History and psychology confirm that it is easier to kill from a distance or from behind, without seeing a person's face or recognizing them as an individual. (This principle applies to crime as well.)

However, Alexander refined his tactics. Instead of allowing his men to break and flee, he maintained disciplined formations even in retreat. His troops never turned their backs and ran.

Instead, they withdrew in an organized manner, still facing the enemy and minimizing casualties.

Alexander perfected the Macedonian Phalanx, an infantry formation some credit to his father, Philip II, while others trace it back to the Sumerians. The phalanx consisted of tightly packed, heavily armed infantry standing shoulder to shoulder in deep ranks. This formation advanced as a unit and, crucially, also retreated as a unit, never scattering or exposing their backs. This disciplined withdrawal method has proven effective in both war and crime prevention throughout history.

Simply turning and running away may seem like good advice, but it often makes you an easier target. It can trigger a chase instinct in criminals or enemy soldiers, increasing your chances of being caught or killed. A smarter approach is a controlled, tactical withdrawal.

Turning your back whether in a combat scenario, self-defense situation, or crime encounter may expose you to extreme danger. Martial artists understand this well. "Giving up your back" is an invitation to be choked, jumped, assaulted, or killed. Instead, back away first to create a safe distance before turning to leave.

That's the best universal advice for an orderly retreat. Everything else is situational. There's no one-size-fits-all answer. After assessing the who, what, where, when, why, and how of your retreat, you can determine the best course of action.

- Back away, still facing the opposition with de-escalation words? That may work sometimes.
- Back away, still facing the opposition saying nothing? That may work sometimes.
- Back away, still facing the opposition with threatening words? That may work sometimes.
- Pre-emptive strike. Then back away, still facing an enemy, then turn. That may work.
- Warning! Though pre-emptive strikes are a great survival strategy, witnesses may report

that you started the fight by "punching" first. You do whatever you want, what you think you need to do, I'm just saying be prepared to explain.
- You are thrown down. You get up, then pick one of the choices above. That may work sometimes.
- You beat the holy hell out of the opposition. Then pick one of the choices as to, if and when to leave. That may work sometimes. You draw. a weapon. The presentation of a knife or a gun has a good success rate of freezing the opposition. How about any expedient weapon? Then pick one of the choices above to leave. That may work
- Pick one among the choices above to leave, like backing away, then turning and run when at a "safe" distance. That may work sometimes.

Sometimes, like vs. an active shooter, you just run.

I mention my old friend and advisor Colonel "Hack" Hackworth, vet of WW II, Korea and Vietnam, and at one time, the most decorated U.S. Army soldier, always had a "go to hell," plan, for when things went to hell and that plan always included the best escape under the worst circumstances. He told me:

> "Hock, sometimes you gotta' blow the horn, (the horn being the trumpet of retreat). Always have a go-to hell plan, and another one when that one goes to hell too."

"For he that fights and runs away, may live to fight another day." This is attributed to Demosthenes, an epic Greek orator. Sometimes even heroes with the most hardcore, "never say die," mottos are smart to retreat. Every stand-off, showdown and ambush is different. There is no one equation to retreat. It's all situational. All we can suggest is that if you can, if not escaping an active shooter or a bomb, you conduct an orderly, smart retreat should you decide to retreat. Face the person with words and command presence, a knife, or a stick, or a gun and back away. Once a considerable distance is achieved then you might turn and leave. Sometimes with a weapon, you might even successfully order the aggressor to leave, and he might comply under threat.

There is no one way to prescribe any one universal orderly retreat, but it is important to understand the concept, teach the idea, and develop and practice some real, "go to hell," plans.

Lions, and tigers and bears! Related escape, somewhat, but about survival and similarities. In my Dallas-Fort Worth Metroplex there was a surprise sighting of a mountain lion. This caused news and reports of, "what to do when a mountain lion confronts you." The US Parks Department advised this orderly retreat:

- Don't run: Running may trigger a "mountain lion's instinct" to chase. Back away.
- Don't crouch or bend over: This can make you look like prey.
- Make noise: Speak loudly and firmly and wave your arms to appear larger. Maintain eye contact: Keep your eyes on the mountain lion until it leaves. (Experts claim that bears do not maintain eye contact. Bears too are now appearing more and more in populated areas.)
- Give them space: Provide an escape route for the mountain lion.
- Pick up children carefully: If you have children with you, pick them up without turning away or bending over.
- Throw objects: If the mountain lion gets closer, you can throw objects in its direction.
- Fight back if attacked: If you are attacked, use rocks, jackets, or sticks to fight back.
- Report the encounter: After the encounter, report it to the local Fish and Wildlife Office or Ranger District.
- Note: I have to state, notice any abstract similarities between this the bad guys. A few, just a few. And for the record, big picture, there are not many animals (or shark) attacks each year.

WHAT Question 6: What is a citizen's arrest?

A frequent question comes up when training citizens, "What if when defending myself and I want to detain, to arrest the attacker for the police?"

In the United States, a private person may arrest another without a warrant for a crime occurring in their presence. However, the crimes for which this is permitted vary by state. Cornell Law School explains and defines this:

> "Citizens' arrest is an arrest made by a private citizen, in contrast to the typical arrest made by a police officer. Citizens' arrests are lawful in certain limited situations, such as when a private citizen personally witnesses a violent crime and then detains the perpetrator.

For example, in tort law, a citizen's arrest is something that any person can do without being held liable for interfering with another person's interests when that interference would otherwise constitute assault, battery, and false imprisonment. This means that any person can physically detain another in order to arrest them, but state statutes define the limited circumstances in which his deprivation of liberty is allowed. In general, the ability to perform a citizen's arrest is the same for a regular person as it is for a police officer without a warrant."

Canadian law permits citizen's arrest and offer these universal tips-advice. "If you do decide to make a citizen's arrest, you should:

- Tell the suspect plainly that you are making a citizen's arrest and that you are holding him or her until police arrive.
- Call the police. (The arrival might take a while. Are you prepared for this wait?)
- Ask explicitly for his or her cooperation until police arrive. Avoid using force if at all possible, and
  use it to the minimum possible otherwise.
- When the police answer the phone and/or arrive, state the plain facts of what happened.
- No matter where you live, check your local laws on this subject."

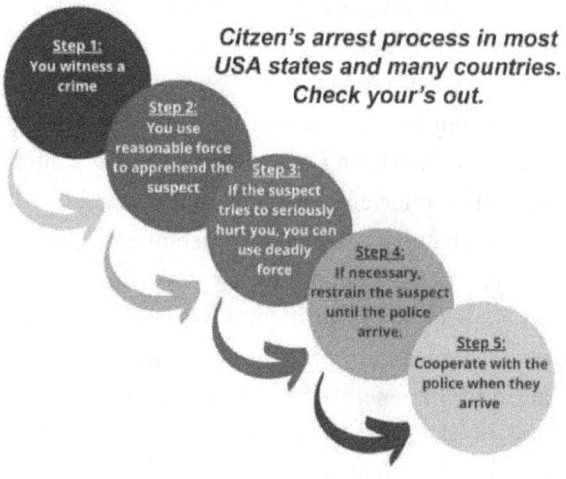

Citizen's arrest process in most USA states and many countries. Check your's out.

Step 1: You witness a crime
Step 2: You use reasonable force to apprehend the suspect
Step 3: If the suspect tries to seriously hurt you, you can use deadly force
Step 4: If necessary, restrain the suspect until the police arrive.
Step 5: Cooperate with the police when they arrive

In Europe, citizen's arrest laws vary by country, but generally allow individuals to detain a suspected criminal if they witness a serious crime being committed, provided they use reasonable force and inform the suspect of the arrest, with the primary focus on preventing further crimes and injuries. Germany and France further define "aid to others in immediate danger" for such actions.

WHAT Question 7: What clothes do you train in?
What clothes will you be fighting in? Abstractum reducere!" (Reduce the abstract.) Maybe you'll get into an argument. Maybe you'll have to make an arrest. Maybe you'll need to protect someone. Or perhaps you're ordered into battle. Maybe you're jumped or ambushed. What will you be wearing? I've written extensively on this subject, but I'll keep it brief here: train in the clothes you expect to fight in. Reduce the abstract.

You don't train for football wearing a baseball uniform. And vice versa. The same for fight training.

Sport outfits are for your hobby. Artsy outfits are for your art. Pursue your happiness. But for survival and mission readiness? Wear what you actually wear in life.

In the field. Civilians should train in civilian clothes. Police officers should train in their uniforms.

For example, it makes no sense for cops learning Brazilian Jiu-Jitsu to wear traditional gis. That's off-mission. Or to train half-prepared, half-dressed. Over the years, I've taught police forces as far away as Australia and as close as the local Texas police academy. More times than I can count, officers showed up in a t-shirt, sweatpants, sneakers, and, at best, their duty gun belt. They should be training in full uniform, just as they would wear on duty. The same goes for military combatives....all this wrestling without gear (and no ground n' pound)? I understand the need to start "loose" with the basics, but let's get to fight training in full gear.

In reality, you won't be barefoot, wearing a gi, a rash guard, superhero suit, or baggy pajamas when a crime or war unfolds. There won't be a mat to cushion your falls. Instead, you'll be burdened by your clothing and mandatory gear. Your wrists won't be wrapped. You won't have a mouthguard. Your eyeballs will be exposed.

One downside is practicing barefoot. Some old-timers claimed it was to "toughen the feet." That may have made sense centuries ago when people were barefoot daily, but today? No. Nowadays, it's mostly about preserving the mats. That's fine, I get it. But maybe you don't need mats at all, or maybe just some of the time?

My old school in Texas had a wooden floor. I understand that hard floors make repetition training and ground fighting tougher, but that's the reality of a fight. BJJ expert Chris Haueter has a great lecture on the evolution and distortions of mats in martial history and ground fighting, that I wish everyone could see. There's a video online if you can find it.

This "sometimes," ties into what I call the "Duration Principle." Strike bare-knuckle on a bag for a while, but when amassing reps? You've done your time, put on MMA gloves. (Not boxing gloves unless you're a boxer.) Training ground-and-pound or catch wrastlin' on a real floor? Start without gear, but when the reps pile up? Slip on the knee and elbow pads, or pull out the mat and finish up. You've earned it. *The Duration Principle.*

In a real fight, you'll likely be wearing shoes and fighting on tile, wood, carpet, concrete, asphalt, stairs, hillsides, the insides and outsides of rural, suburban and urban worlds. Train in the clothes and gear you'll actually be fighting in...and on, as much as possible. If you are just training for art and hobby? Then have fun. Just don't be schizophrenic about it. "Abstractum reducere!" (Reduce the abstract.)

WHAT Question 8: What training should you prioritize? What You NEED vs. What You WANT. What do you think you want versus what you really need? Sports? A hobby? Desperate self-defense? Combatives? Most people don't know the difference. Their "martial I.Q." is low. They're either beginners or have a limited scope of understanding.

We've already discussed in past essays how to choose the right trainer or instructor and what subjects should be prioritized in training. If you're skipping sports and traditional martial arts, as I did long ago,

QUADFECTA MARTIAL STUDIES

"WELCOME TO THE FOURTH DIMENSION"

....KICKBOXING - GROUND N' POUND - WRESTLING - WEAPONS.......

hopefully, you'll find a dedicated, generic, "clean" doctrine-based combatives instructor rather than someone bound by dogma. You've probably heard of the Trifecta (three-bees), but how about the QUADfecta (four-bees)?

Yes, it exists in the Fourth Dimension, no big deal according to Dr. Strange (and, these days, numerous brilliant scientists alerting the world to other dimensions).

When push comes to shove, reality boils down to fighting three main types of adversaries: criminals, enemy soldiers, and our "drunk uncles" (a term I use for troublesome friends and relatives we can't just bust up, break, or kill).

The Martial Quadfecta. In researching solutions with a keen eye for "use of force," I explore what I call the Martial Quadfecta:

1) Kickboxing. Filter for reality.
2) Ground n' Pound. Filter for reality.
3) Wrestling (I prefer Catch Wrestling over BJJ for controlling your drunk uncles, family, and friends. Filter for reality.
4) Weapons (modern sticks, knives, guns, not ancient artifacts). Filter for reality.

Entering the fourth dimension. Any one of these fighting systems alone is one-dimensional. Combining two makes it two-dimensional. Three? Three-dimensional. But in today's world, the Multiverse, as Dr. Strange suggests, we must think in FOUR dimensions. Enter the Quadfecta.

What techniques can I adopt from these four sources for crime and warfare? Some say "steal" from these sources; others say "take," "co-opt," "borrow," or "adopt." Whatever term you prefer, I am always on the hunt, while keeping things simple, more like checkers than chess. Every choice is run through the essential questions: Who, What, Where, When, How, and Why.

Decisions about training should be made by people with a high "4-D Martial I.Q.," meaning they're smart enough to make informed choices. I've seen many tough combatives instructors casually preach "slay, blind, destroy" as if every opponent were an enemy Nazi soldier. But reality isn't that black and white. The wrong use of force at the wrong time on the wrong person can land you in jail. Not everyone is a Nazi. Stay in Your Lane. If you're teaching a one, two, or three-dimensional martial art, be careful about advertising it as "ultimate self-defense" or "combatives." It's not. You're merely tiptoeing around the Quadfecta.

That's fine as many people enjoy staying in their one, two, or three-dimensional martial arts lanes. If that's you, great. I'm happy if you're happy. Just be aware of your limitations. Understand where your training fits in the bigger picture. Enjoy your hobby, your exercise, your camaraderie. Getting off the couch is a victory in itself.

But don't try to sell a police officer or military veteran on the idea that you're teaching real, life-or-death combatives. We won't buy it. Crime and war change your perspective. Practitioners seeking genuine self-defense should learn from vets or those certified by them. Even then, we're all works in progress, just some further down the line. (I wonder if anyone will ever call their school "Quadfecta Martial Arts." Go for it! I haven't copyrighted it, and I won't. I just love the idea of spreading "Quadfecterism.")

WHAT Question 9: What Probabilities? What possibilities? There are two groups of problem areas to think about. Probabilities and possibilities. Real world fighters in crime and war start with probabilities as a priority. But then you cannot ignore possibilities. I am going to slightly paraphrase Sherlock Holmes here (well, you know, actually Sir Arthur Conan Doyle): "When you have eliminated all which is possible, then whatever remains, however improbable, must be the truth."

The "What If Factor" will drive you crazy in thought experiments. Instructors know this too well with frisky students, dreaming up all the "what ifs."

Reduce the insanity by listing probabilities first versus possibilities second, as investigated by the Ws and H questions. Upon investigation, work on those small and big probabilities. Then, explore the lesser uncommon possibilities, so your fortunes can favor the prepared. Please don't do this in reverse!

WHAT Question 10: What can-should you do and what can't-shouldn't you do? Every competent person and organization either have, or should have a well-defined mission statement. It serves as the foundation. In many operations, particularly in a training company like mine, the mission statement guides doctrine, prevents dogma, and helps avoid confusion and even innocent hypocrisy. Organizations and individuals must clearly define what they do. One way to strengthen this definition is by also clarifying what they don't do.

The Consequences of a Mission Statement. Every mission statement carries consequences, both intended and unintended. While a well-planned mission can yield great results, unforeseen challenges will always arise. Being flexible allows you to adapt to changes while staying true to your mission. Recognizing what you are not doesn't limit you; it keeps you grounded, realistic, and mission-focused. The reasons behind saying no to certain things, the why, could take a few lines, a paragraph, or even a chapter to explain. However, for the sake of brevity, this essay won't delve into such

details. Instead, here are a few key considerations using the classic Ws and H:
- WHO needs your mission?
- WHAT is your mission? What goal?
- WHERE is your mission needed?
- WHEN is it needed? Is time running out?
- HOW will you implement your mission? How will you spread the word? How crucial is your message?
- WHY does your mission matter?

A strong mission statement provides clarity, direction, and purpose. It defines what you stand for and what you don't, keeping you focused and effective.

WHAT Question 11: What If and What's it gonna' take games. I have already mentioned the importance of crisis rehearsal. That's playing the what if such-and-such happens, what will I do?

As a patrolman and detective, I've arrested approximately 1,000 people over 26 years. I often had to quickly assess how difficult an arrest might be. How well could the suspect fight? Sometimes, I was surprised, both in good and bad ways.

To stay sharp, I would mentally evaluate the fighting capabilities of various "innocent" people. What would it take to win against this or that random opponent?

I played this guessing and forecasting game while shopping or attending events, making quick assessments. Of course, assessments can be wrong.

You might look at a giant, obese man and assume, "He's out of shape, an easy fight." Then you discover he's a defensive lineman for the Pittsburgh Steelers. Oops. But it certainly made trips to the supermarket more interesting.

You! Yeah you! You're under arrest! (Gulp....)

What Question 12: What Would Jesus do? The Sam Elliot Decision. The "Sam Elliott Decision": To Treat or Not to Treat?

Years ago, I watched a western movie starring Sam Elliott. I can't recall the title, but the plot stuck with me. Two men came to kill him at a remote cabin. He shot them. One survived. Without hesitation, Elliott's character hauled the wounded man inside and began treating him. From that moment, in my mind, I had a name for this rescue, critical situation, the "Sam Elliott Decision."

Today, serious gun owners and professionals spend time on tactical medicine. But for whom? Themselves, their family, their colleagues, but what about the suspect? Should you move in and treat the bad guy? Should you even check if he's still a threat? Or do you run, hide and call 911?

For we Christians, Jesus remains the highest moral standard. When faced with challenges, one might pause and contemplate: What would Jesus do? While we are not Jesus and may not always be able to follow through in this messy, fallen world, it is still a valuable lofty, moral and ethical mental exercise to consider.

This scenario forces us to consider an aftermath that all gun owners, particularly those in law enforcement or the military, need to think about. What do you do with a freshly wounded opponent?

The Amber Guyger Case. A lesson in aftermath.

Consider the case of former Dallas police officer Amber Guyger. She returned home late at night after a 12-hour shift, entered what she mistakenly believed to be her apartment, and fatally shot a man, a neighbor, watching TV in his own living room. A white female officer killing a black man in his own home was national and international news. The situation was tragic, and the local community erupted in controversy and protests.

At her trial, Guyger testified, an unusual move. Under prosecution questioning, it was revealed that although she had medical supplies in her police backpack, she made no effort to render aid. She called 911 and otherwise wandered in and out of the apartment, talking on her phone. In court, this medical inaction suggested a lack of care, reinforcing accusations of racial bias and contributing to her conviction. She was sentenced to ten years in prison.

Historically, law enforcement receives only rudimentary first-aid training. The general practice was to secure weapons, handcuff the suspect, and call an ambulance. I followed this procedure numerous times. In one case in the 1980s, we shot an armed robber. I don't recall if he was dead, but we handcuffed him and waited for the ambulance. Neighbors accused us of letting him die. His mother sued the department, and though it was Texas some 35 years ago, the city settled for quite a substantial sum. Frankly, none of us even considered treating him.

This "no-treat" approach became highly scrutinized after the infamous North Hollywood shootout in the 1990s. The police allowed a wounded robber to bleed out while standing around. The robber's family sued LAPD for neglecting his medical care. Though public sympathy was minimal, LAPD also settled for a significant amount.

In 1989, the U.S. Supreme Court established that police do not have a general constitutional duty to protect or provide aid unless a person is in their custody. Once someone is in custody (handcuffed, restrained, or otherwise detained), officers generally owe a duty of care under the Constitution's Due Process Clause. This includes providing or arranging medical treatment if someone is injured. If it's safe enough, this the definition of "secure the scene."

The issue of post-shooting medical care is not limited to police officers. Consider the case of Navy SEAL Eddie Gallagher, accused of war crimes, including the alleged execution of a wounded teenage combatant. One of the accusations against him was failing to provide adequate medical treatment. The implications were significant.

My question is, how long before this becomes a citizen problem? I think it is already to some extent. And probably growing. Shooting a criminal and refusing to help may seem justified in the heat of the moment, "He was trying to kill me, so screw him." But while that sentiment might fly in a barroom conversation, it might not hold up in criminal and civil court.

Recently respected former officer and renown gun expert Massad Ayoob advised citizens NOT treat freshly wounded criminals. He basically said it was unsafe. Don't do it. But what if it was safe? What if juries and courts thought it was safe? You best think about that.

The growing legal landscape. More law enforcement agencies are now mandating medical follow-ups for those shot after a shooting. Since I first wrote about this in 2018, I have seen numerous body cam videos of police officers immediately administering aid after shooting a suspect. They must.

The legal concept of "failure to render aid" currently applies mainly to traffic accidents. However, it has long been used in civil courts, and I fear it will eventually creep into criminal law, especially in left-leaning jurisdictions.

The Kyle Rittenhouse case: A tactical and legal strategy.

During his trial, Kyle Rittenhouse testified that he wanted to check on one of the men he shot to render aid but couldn't because of the hostile crowd. The jury heard this. His statement countered the prosecution's claim of malicious intent and likely contributed to his acquittal. His desire took the sting out of the parts of charges.

Minnesota now has a statute that imposes a "duty to render aid" on any person who discharges a firearm and knows or should know that the discharge caused bodily harm.

"A person who discharges a firearm and knows or has reason to know that the discharge has caused bodily harm ... shall ... render immediate reason able assistance to the injured person."

It also provides that it is a crime to violate this duty (criminal penalties vary by severity of harm). But there is a statutory defense - if the person reasonably perceives that rendering aid would pose a "significant risk of bodily harm" to themselves or others, then they may be justified not to do so.

Even if no criminal charges are filed, the family of the attacker could bring a wrongful death lawsuit. Any given jury may look negatively upon a shooter who made no effort to help the attacker survive, especially after the danger passed.

If prosecutors or a civil attorneys question your decision, you need a logical, well-reasoned answer about why you did or didn't. You must articulate why you did or did not. There are plenty of possible, logical reasons.

I think the answers are situational. If you shoot someone, if even Hannibal Lecter himself, someone in the public and the legal system will scrutinize whether you could or could not try to save the life. Often that will be a civil jury. The Sam Elliott Decision is a real-world dilemma that requires thought, preparation, and articulation.

WHAT Question 13: What is the Right Reason?

I recall the old police tale of a rookie asking his police chief about what to do on his first day and then, his whole career? The chief's answer was, "Well, always do the right thing."

Of course, you need training to discover the right thing in some situations. But I wish for you to consider a mighty important WHAT question. What is the right reason? The best motive, real reason you do what you should do.

Have you selected your course, your forecast, your decision for all the best, right reasons? The righteous reason? There could be a whole book called *The Right Reason*. But you should write that book for yourself.

WHAT Question 14: What is your perception, your preconceived idea of your first or next fight?
(Or, how we learned to wrestle with our preconceived notions about fighting.)

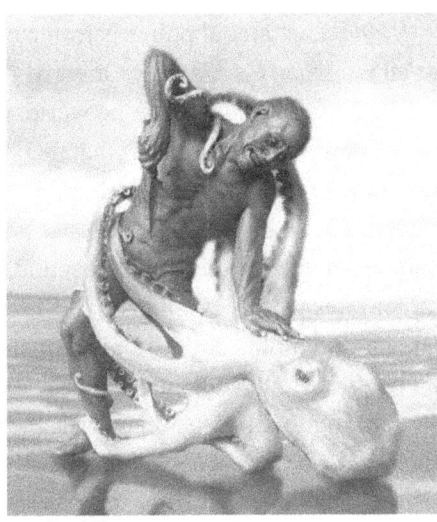

Is your idea of your first or next fight as preposterous as this? Might be?

Oftentimes, what you imagine your next fight will be... isn't what it will actually be. This is, by far, one of the must important essays I've ever written.

Feeling nostalgic one night, I recently watched the first episode of the 1980s T.J. Hooker cop show. I was already a street cop and detective for years when this was on prime-time TV. On patrol in a 1970s, giant squad car, prowling residential streets, Hooker and his rookie partner drove through a Los Angeles neighborhood as Hooker lectured the rookie about the shame and horror of how people cowered and hid in their houses, fearing the crime on the L.A. streets. That was 1981!

"They" were already scaring the bejesus out of people back then. Of course, that was just TV drama, but these fear-driven narratives fed and continue to feed people today. Perceptions shape reality.

Defining perceptions. I am an old police patrolman and detective from an era when Community Oriented Policing (C.O.P.) was supposed to save the world. One of the main points of this movement was that the perception of crime, the idea of it, was just as real to citizens as actual crime data.

The disconnect between reality and perception, look at how the murder rates in very small parts of Chicago, Baltimore, Oakland, Memphis, or St. Louis influence the opinions of people within and outside those cities, states, and even the country. In other countries, these tiny crime-ridden jurisdictions shape global perceptions of the USA. People from elsewhere consider these entire cities as crime hellholes when, in reality, they are not. The big picture tells a different story.

Usually, the perception of crime is far greater than the actual crime. So, we police with C.O.P. not only had to fight real crime but also had to engage in an advertising and public relations campaign against the perception of crime. Fighting fake crime. We also had to make people... well, happy.

The reality of big picture crime: most people in the USA and many other civilized countries will never be victims of crime. That's a statistical fact. But that doesn't mean people should ignore crime. They should still have some realistic, healthy fears, perceptions, and plans. What do they perceive? What do they imagine? A home invader? Rapist? Mugger? Mass shooter? Crazy guy? Serial killer? Kidnapping? Bar fight? Road rage? A sword fight? A 28-inch stick fight?

In the same way, some even have a specific perception of how they will fight. Gun? Knife? MMA? WWII combatives? Kill? Maim? Contain? Negotiate? Pray? It certainly helps if those perceptions are as accurate as possible. Perception, as defined by "Professor Webster:"

> "A way of regarding, trying to understand, or interpreting something; a mental impression."

<u>*The power of mental impressions.*</u> How deep is that paranoid perception of criminals? Has it changed over time? Aside from breaking news "alerts," many perceptions about fighting bad guys are subliminally shaped by books, movies, TV, and even personal fantasies.

The same applies to fights. Remember back in the 1970s–1990s when Chuck Norris or Jean-Claude Van Damme would kick a bad guy down? The bad guy would crash, and the Chucks and Claudes would just stand there, striking a heroic pose, maybe bouncing up and down, waiting for the serial killer or hitman to stand back up and continue the classic fight.

Art imitates life, and life mimics art. How many people have actually waited for bad guys to stand back up after they have been knocked down? That was the pop culture movie fight. Nobody dove in with fists flying like real people do, like the ground-and-pound UFC fighters do now.

Sure, some fights start with stand-off posing, and, as we used to call it in the Army military police, pre-fight "profiling." Taking off a wristwatch. Removing a shirt and spreading out their back in an animalistic display. And within some street fights, there are breakaways where some boxing-like or kickboxing-like exchanges might happen.

The illusion of preparedness. We had a champion black belt in our classical karate school decades ago who got into a bar fight, and lost. He came to class and told the school owner: *"I was in a fight last night, and it wasn't anything like I thought it would be."*

If you are in a non-sport self-defense class, your student should return from a fight and say, *"I was in a fight last night, and it was just like you told me."*

Proper perception. Perception is the foundation of training, isn't it? We, martial artists, civilians, police, and military, train based on our perception of what our next fight will be like. If you are sport fighting, you know exactly who, what, where, when, how, and why about your scheduled fight. Even as a soldier, you have solid intelligence about what might happen. But what about sporadic, unpredictable encounters with criminals? What about that first-time fight for the average citizen?

Many modern fighting systems inadvertently train for a fight that happens only in a ring, in a bar, or in the parking lot outside a bar. Meanwhile, a soldier in Syria or a cop responding to a factory disturbance has entirely different scenarios in mind. What about dealing with a rapist atop a woman in a bedroom or on a couch?

People often train for what they perceive their next fight will be like, based on the classes they take. And often, they're training for the wrong fight.

The stupid fighter. Another problem is the novice fighter or, as I like to call him, the stupid fighter. Statistically, your first or next opponent is likely to be an untrained idiot. You've spent years training to fight skilled opponents who mirror your own techniques. Then idiot-boy walks up, spits in your eye, and smashes a chair over your head. Mark Twain put it so well:

> "The best swordsman in the world doesn't need to fear the second-best swordsman. No, the person he should fear is the ignorant antagonist who has never held a sword before. The expert isn't prepared for him."

The reality of fights. Real fights happen everywhere, inside homes full of furniture, in parking lots, on streets, in the mud, in the rain. One time, a crazy guy and I slid down a long, slick, muddy hill in heavy rain, fists flying, outside a hospital on a hill. No mats. No clean footwork. Just chaos.

People fight in rural, urban, and suburban areas, inside and outside buildings, at any time of day, in any weather.

When we prepare, we perceive. It's wise to forecast, to analyze crime data, and to recognize that, statistically, most people will never be crime victims. But that doesn't mean we should be complacent.

At least now, with the advent of reality fight videos on YouTube, we have a more realistic perception of the chaos of a fight, not just the Hollywood version. We should learn and train for fundamental, adaptable skills that apply to any situation.

WHAT Question 15: What (if any) Weapons? Should You Carry and Use One? The next Q&A series raises an important question: Should you carry a weapon? Would you? Could you? Do you have the resolve to use a stick, knife, or gun if needed? What are the weapon laws where you live, work, and travel? Do you truly need a weapon?

Years ago, a widely shared statistic claimed that "40% of the time when you fight someone, they are armed." A concerning thought, though no clear source backed it up. Later, a retired state police supervisor conducted a study and concluded that 90% of people involved in physical altercations were armed! That's even more alarming. However, I haven't listed his name here, as his once-popular police training course has since faded from public view, and his study is no longer available online. Still, this statistic was widely circulated. Is it accurate? I found the 90% figure hard to

YOU HAVE A MORAL, ETHICAL AND LEGAL RESPONSIBILITY WHEN YOU CARRY AND CHOSE TO USE A WEAPON.

believe, but perhaps you've heard similar numbers. The study is gone now.

The point is: Think critically. Make an informed decision about weapons. Be prepared to justify your reasoning. In law enforcement, we are trained to fight unarmed, even while carrying weapons like a baton, knife, pepper spray, stun gun, or firearms. Choosing the right tools at the right time requires discipline, something that can be developed through proper training.

What weapon? Your body. Your unarmed self is a weapon in its own right. Have you decided to navigate life, both at home and away, without carrying any weapons? Do you live somewhere that prohibits them entirely? (So that only outlaws have them?) What weapons might a potential attacker have? If you're unarmed, you may have to fight both unarmed and armed opponents. Training is required.

What weapon? Stick / impact weapon. Are you the type of person who carries an impact weapon daily? This question ties back to self-awareness: Are you confident? Ignorant? Do you live in an area where all weapons, including impact weapons are illegal?

Consider what's already within reach. Do you keep a stick, bat, or heavy tool in your home, car, or garage? Even everyday objects, like a large ice scraper or baseball bat, could serve as makeshift weapons.

However, be mindful of perceptions, carrying a bat for protection can raise suspicion. If you must carry one, say, in your car, consider attaching a ball glove to create the appearance of legitimate "sports carry."

Think critically about impact weapons: axe handles, batons, expandable sticks. What might an attacker have? As someone with multiple black belts in Filipino Martial Arts (FMA), I've spent countless hours training with sticks. But realistically, how often will you find yourself in a duel where both parties coincidentally wield 28-inch sticks? Training with impact weapons has benefits like fitness, discipline, skill-building, but real-world survival demands practical decision-making. Make an informed choice. Training is essential.

What weapon? Knife. Are you someone who carries a knife every day? The presence and use of a blade carry legal and social stigmas. A simple pocket knife is common, but a tactical-looking knife can change how you're perceived in a legal situation. Be mindful of the look and the branding. Would you rather explain owning a tool labeled "Plumber's Helper" or one called "Seal Team Six Combat Throat Slitter"?

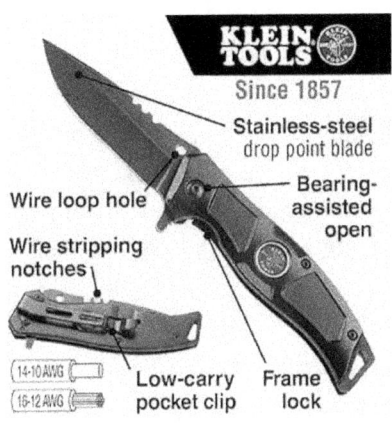

For example, consider this widely available knife from Home Depot: Klein Tools 44228 Electrician's Pocket Knife Stainless steel blade, perfect for splitting cable, stripping wire, and everyday tasks. A practical tool like this costs under $20 and has a clear purpose. Contrast that with a curved karambit or a combat knife, which may be harder to justify in court. Also, be aware of how your overall image influences perception, visible tattoos with violent themes could be used against you in legal proceedings.

Think carefully and make an informed choice about carrying a knife. Training is required.

What weapon? Firearm? Who are you to carry a gun? Why would you? Should you navigate life, both at home and away, with or without one? The decision to carry a firearm comes with significant responsibility and legal considerations. As with knives, branding matters. Would you rather carry a "Peacemaker" or a "Widow Maker"? What weapons might a potential attacker have? Consider all factors and make a well-reasoned decision about firearms. Training is required.

Weapon selection. Whether none...or some, or many should be an informed and deliberate decision. Think critically, assess your needs, and train accordingly. Your choices must align with the law, your personal philosophy, and your ability to act responsibly.

Where I live I can't have a gun or knife, so why should I other... I teach in some 8-10 countries a year, so I am in highly restricted weapon jurisdictions. Lest of all some very liberal states in the USA. Attendees tell me,

"Why should I bother training anything with a pistol or a knife, I can't carry one. So, all we can do is unarmed." The conversation begins.

"Do criminals here have guns and knives?" I ask.

"Yes, they do."

"In your unarmed courses, do you practice weapon disarming?" I ask.

"Ooooh yes we do. I do Ballistic-Galactic Krav Maga! We do 43 disarms."

"Okay, so, you disarm a knife and a pistol. Two things happen. The weapon is either on the ground, or in your hand. One is now in your hands! Do you know what do with it? What end the bullet comes out? A safety? How to best use a pistol or knife?"

Citizen dumbfounded...

"You may not hold a weapon you carry, but you may suddenly hold a criminal's weapon. And you need to use it. Fast."

Citizen dumbfounded...

"You know knives are everywhere. Think about being in a restaurant. or your house. Most knife attacks in the civilized world are done with kitchen knives."

Dumfounded...

If you are studying real world survival, not a hobby or art, it is a hand, impact weapon-stick, knife, gun, crime and war world. It behooves a survivalist to know their way in, out and around weapons, regardless of the local carry laws.

WHAT Question 16: Speaking of weapons, what is "weapon brandishing?" If you do show or pull a weapon out in some very liberal, anti-weapon-carry localities, even if to scare off an attacker, you might suffer from weapon brandishing charges-laws. Is just showing a weapon sheer brandishing? When is it? When is it not? U. S. Federal law defines brandished as:

> "...with reference to a dangerous weapon (including a firearm) means that all or part of the weapon was displayed, or the presence of the weapon was otherwise made known to another person, in order to intimidate that person, regardless of whether the weapon was directly visible to that person. Accordingly, although the dangerous weapon does not have to be directly visible, the weapon must be present." - (18 USCS Appx § 1B1.1)

In Canada, a weapon is referred to in legalese as an "object." So, one must do a dog-and-pony show on what "object" was used in the situation. Pencil? Screwdriver? Toothpick? Potato chip? Thumb?

(Thumb? Yes, there's always some wacky place that reclassifies unarmed tactics as a "weapon."). The US Carry webpage says,
- Example 1: Brandishing a weapon can be called a lot of different things in different states.
- Example 2: Improper Exhibition of a Weapon.
- Example 3: Defensive Display.
- Example 4: Unlawful Display.

Retired special operations Ben Findley advises, "'brandishing' or 'improper exhibition' or 'defensive display' or 'unlawful display' (or whatever your state and jurisdiction call it) depends specifically on your geography. Very generally, however, for an operating definition 'brandishing' means to display, show, wave, or exhibit the firearm in a manner which another person might find threatening."

You can see how different this can be subjectively interpreted by different "reasonable" individuals and entities. The crime can actually be committed in some states by not even pointing a firearm at someone. In some states it's a misdemeanor crime and in others a felony. So, focus, think rationally, know your city, county, state and country law, and be careful out there.

A quick review of several countries will reveal that weapon brandishing laws include odd words as legal terms like:

-rude, (was the gun-toter obnoxious and rude?)
-careless (was the knife-toter waving it around?)
-angry, (was the stick-toter yelling and red-faced?)
-threatening manner? For brandishing? What? For many, the whole point of aiming a stick, knife and

gun at a brewing bad guy is to be threatening! What then is the line between a smart preemptive strike, a smart weapon show or pull and a crime? How can we make it all become justified self-defense?

As a cop for three decades, I am alive today because I pulled my gun out a number of times, just before I REALLY needed it. I use the line "the fastest quick draw is getting your gun just before you really needed it," which borderlines our brandishing subject. Be justified in your reasoning.

The remarkable researcher, legal expert and police vet Massad Ayoob says, "When an unidentifiable citizen clears leather without obvious reason, folks start screaming and calling 9-1-1, and words like "brandishing" start being uttered. Thus, circumstances often constrain the law-abiding armed citizen from drawing until the danger is more apparent, which usually means the danger is greater. Therefore, often having to wait longer to reach for the gun, the armed citizen may actually need quick-draw skills more than the law enforcement officer."

> A. Nathan Zeliff, a California attorney reports, "Brandishing – drawing your firearm pursuant to a lawful act of self-defense should not be considered brandishing. However, if it is determined that you drew your firearm and the facts and circumstances show that you drew or exhibited the firearm in a threatening manner, and that such was not in self-defense or in defense of another, then you may face charges of brandishing."

I am not so sure this brandishing topic comes up all that much in weapon courses other than gun material? Or certainly not enough. Here's some collective words of wisdom on the subject., advice like this:

- Prepare for problems by using the Who, What, Where, When, How and Why questions.
- Avoid possible dangerous arguments and confrontations when possible. Conduct yourself with smart, self-control. Leave if you morally, ethically can.
- Obtain a valid, concealed carry license for all your weapons.
- Keep your weapon concealed. Do not open carry it. You might get in trouble for opening your jacket and displaying your weapon, in some very liberal jurisdictions.
- Do not display a stick/baton, knife or pistol, or threaten deadly force unless you, or others are threatened with imminent death or serious, bodily harm.
- Attend "fundamentals of fighting" classes with and without weapons training and learn the use of deadly force laws in your city, county, state and country.
- Have a logical, reasonable motive to pull the weapon.
- I can say from experience and lessons that if you must point a weapon at someone to stop them, many will routinely yell out, "Oh, what are you gonna' do? Shoot me? (or stab me? or strike me?)" One good answer might be "I

don't know yet. Depends on what you do."
Think about your answer to this ever-so-common remark.

And furthermore, be the first to call. Massad Ayoob recommends the following steps to take when calling 911 after a self-defense situation, shooting or otherwise.
- Be the first to call. Be the victim-complainant.
- Identify yourself: State your name.
- State the situation: Briefly explain what happened.
- Provide a description: Give a quick description of yourself and the bad guys.
- Give your location: Clearly state where you are.
- Request assistance: Ask for both police and an ambulance (if needed).
- Conclude the call: After providing the necessary information, conclude the call and hang up. (Ayoob recommends not staying on the line with 911 so you can focus on your safety and avoid making any statements that could incriminate you.)

WHAT Question 17: What happens to your body?
What happens to you when you start to fight? Lots, and genetically. This remarkable book by Robert Sapolsky is the best, a gamechanger on the subject. In the beginning chapters, Sapolsky extensively covers what happens when someone starts to fight, mentally, physiologically and genetically unlike anyone you have

read or heard of before. Pass on all others and read this amazing book.

> "Reading this book as a neuroscientist...Robert Sapolsky is a giant in the field of behavioral neuroscience." - Dr. Helen Collins

What Question 18: What should you do when you are alone and have a heart attack? My wife died at home alone from a sudden heart attack. And, a few times in my career I have had to perform CPR on heart attack victims, so I have a deep connection to this subject.

One time on a woman who was punched so hard in the chest by her husband her heart stopped. Then he left her on the street. I was dispatched to this domestic. A punch to the chest stopping a heart? Yes.

> "*Commotio cordis* is an extremely rare, serious medical condition that can happen after a sudden, blunt impact to the chest. If the physical blow hits during a narrow window in

the heart rhythm, it can disrupt the heartbeat and cause sudden cardiac arrest."

- HeartOrg.com

I apparently saved her life with CPR, replaced by EMTs. (And later arrested her husband.) I report this here because people in the martial training world need to know this could happen, and what to expect when punched in this manner, in the chest. I know of several instances where in baseball, line drives and fast pitches-throws hit players' chest, and some football tackles, and stopped their hearts. And at times, the punch or sports impact set the stage for later heart attacks when one might be alone. Later...is your chest just hurting from the impact? Or...?

To cover this scary subject, here's some definitions, advice and established tips on what to do when you are alone and are suffering a heart attack.

- Chest ache, pain or discomfort. Chest discomfort usually occurs in the center of the chest.
- A feeling of heaviness, tightness, pressure, aching, burning, numbness, fullness, or squeezing.
- The pain can either last for several minutes, go away, or come back repeatedly.
- Pains in the torso/upper body, including the arms, left shoulder, back, neck, jaw, or stomach.
- Difficulty breathing/feeling out of breath
- Sweating or "cold" sweating.
- A sense of fullness, indigestion, or choking.

- Nausea or vomiting, lightheadedness, dizziness, feeling like you might faint.
- Unexplained tiredness, weakness, extreme weakness (like you can't do easy tasks), or severe anxiety reacting to symptoms.
- Rapid or irregular heartbeat.

Alone? What should you do?

- Call 911 The most important thing to do when you suspect a heart attack is to call emergency medical services. Always call 911 before you attempt to contact anyone else. Calling 911 will almost always be the quickest way to get treatment. Even if you live in an area that an ambulance may have difficulty getting to, the emergency dispatcher can provide you with instructions on minimizing the damage.
- Consider contacting someone to come over immediately. If you have a trustworthy neighbor or relative who lives nearby, make another phone call asking that person to come to meet you. Having another person nearby can be helpful if you suddenly go into cardiac arrest. You should only do this if the emergency dispatcher permits you to get off the phone or if you have a second line you can call on while the dispatcher stays on the first line. Do not rely on another person to get you to the hospital. Wait for emergency paramedics to show up.
- Chew on aspirin. Chew and swallow a single 325-mg or two tablets of baby aspirin 81-mg. Chewing on aspirin is especially effective if

done within 30 minutes of your first symptoms. Aspirin inhibits platelet development. Taking aspirin can delay the formation of blood clots that could further block your arteries during a heart attack. Chew the aspirin before swallowing it. By chewing the aspirin, you release more of the medicine directly into your stomach and hasten its ability to get into your bloodstream. Do NOT use this treatment If you are on a medication that interacts poorly with aspirin or have otherwise been told by your doctor not to take aspirin or if you are allergic to aspirin.

- Do not attempt to drive. Driving yourself to the hospital is not recommended. If you begin to experience heart attack symptoms while driving, immediately pull off to the side of the road. If you suffer from a cardiac arrest, you will pass out. This is the main reason why driving while suffering from a heart attack is not advised.

- Remain calm. As frightening as a heart attack is, rushing around or putting yourself into a state of panic can worsen the problem. Relax as much as possible to keep your heart rate steady and calm. Use the tactical breathing method as described later in this book.

- Lie down. Lie on your back and raise your legs upward. This opens up the diaphragm, making it easier for you to breathe and supply oxygen to your blood. Make the position easier to maintain by propping your legs up on pillows

or another object. You could also lie down on the floor with your legs propped up on a couch or chair. Make your way carefully if you cannot immediately lie down or sit, such as working on a ladder or crossing traffic. If you feel dizzy or unable to walk correctly, watching a fixed object such as the horizon or a large, fixed object might help calm you
and help you control the situation until help arrives. Consider lying down in front of an open window, open door, fan, or air conditioner. Providing yourself with a consistent stream of fresh air can help supply your heart with oxygen.

- Do not attempt Cough CPR. A common myth is that you can survive a heart attack alone by coughing in a particular manner. This probably won't work, and worse still, attempting this technique may put you in more danger. Attempting this procedure on your own can cause you to accidentally work against the rhythm of your heart and make it harder to get oxygen into your blood rather than easier.
- Avoid food and drink. Eating and drinking are probably the last things on your mind when you experience a heart attack, but just in case, you should avoid food and drink even if you want them. Having anything other than aspirin in your system can make it more complicated for paramedics to give you adequate treatment. If necessary, you can swallow a little water to

help you get the aspirin into your system, but even this should be avoided if possible.
- Lower Your Body Temperature. If you can lower your body temperature, you increase your chance of surviving a heart attack. A quick way to decrease your body temperature is by placing a cool towel under your arms or on your wrist. Using ice packs is even better if you have a few available.

  According to the American Heart Association, naturally lowering the temperature in this way decreases your risk of experiencing brain damage. It stops chemical reactions that can damage your organs when full oxygen flow resumes. Lowering your body temperature also helps regulate your heart rate and encourages steady breathing.
- Follow-Up: Talk to your doctor about what to do in the future. Suffering a heart attack increases your risk of experiencing a heart attack in the future. When you survive your heart attack this time, you should talk with your physician to discuss improving your chances of survival if you should suffer from one again."

What Question 19: What is Deconflict-Defconfliction? The world has heard of de-escalation, what of this? It's a jargon word in the military and policing. What does deconflict mean? Deconflict means to do something to avoid or remove the potential for conflict. Deconflict is especially used in a military context to refer to an action intended to avoid conflict between non-enemy forces in an area or to remove elements (such as weapons) that could lead to dangerous situations. It can also be used in everyday situations when there is disagreement or things are in opposition to each other. Example statement: "We are in the process of deconflicting this zone by clearing it of all military personnel and equipment."

In law enforcement, deconflict means to identify and resolve overlaps, conflicts, or interference between operations, investigations, or responding units so officers don't accidentally work against each other, or worse, mistake one another for suspects. Plain talk: making sure everyone knows who's doing "who, what, where, when, how and why."

Certainly, in a world of necessary fast responses, security, leaks, spies and mole problems, not everything can be revealed to everybody.

Citizens could certainly make applications to deconflict their lives.

WHAT Summary

By now you surely realize there are many more WHAT questions. Continue asking and answering small and big WHAT questions.

The What Question Review

Who are you within this what category?
What are you within this what category?
Where are you within this what category?
When are you within this what category?
How are you within this what category?
Why are you within this what category?

Chapter 5:
The Where Question and Confrontations

The photo above is one of my favorite staged gun photos, taken from a gun magazine. Clearly, it depicts an all-American dad, mom, and child, all somehow, someway "round back." You know, the ... round back. Behind some place, maybe near a dumpster? How did that happen? How did they end up...round-backed? Lost in the sinister round-back world where skinny, masked men with knives lurk in the shadows, waiting for their all-American prey to stumble by...

It's all about being in the wrong place at the wrong time. But even being in the right place at the right time can go wrong. Where? Where are you going? Where exactly are you? And why are you staying there? You could be in the wrong place at the wrong time, or in the right place at the right time and still find yourself in trouble.

Trouble walks up to you just as easily as you walk up to it. Friends, family, police, and prosecutors will often ask, "Why did you go there?" And then, "Why did you stay?" The key is to choose the least troublesome places and when trouble starts, leave.

Getting there. Places matter. In the book, "The Power of Intuition," *sensemaker* pioneer Gary Klein describes working with the U.S. Marines on decision-making. One study examined a headquarters directive that assumed troops could move at an average pace of 2.5 miles per hour.

This number became the default for mission planning. But those on the ground, NCOs and platoon leaders knew the reality was different. Where are they walking? Desert sands? Swamps? A city? Asphalt roads? Where makes all the difference. Terrain changes time and distance. And coming and going to places on time isn't just a logistical concern. It can be dangerous. Klein's study connected the boots on the ground with the command for these calculations.

Having been part of two presidential security details and having organized protection for high-value VIPs such as for former NYC mayor Rudy Giuliani, I always use the essential Ws and H questions to plan security. Anyone in protective services knows that some of the riskiest moments are arrivals and departures. An inside location can be secured, but the coming and going? Those brief, exposed transitions? Are often more unpredictable. So, I ask, where are you now? Where are you going?

WHERE Question 1: Where Do You Live? In her must-read book, *"The Unthinkable: Who Survives When Disaster Strikes,"* Amanda Ripley points out that over 80% of Americans live in or near cities, relying on sprawling public and private networks for essentials like food, water, power, transportation, medicine, and emergency services. She warns, "We make nothing for ourselves." We buy everything we need.

What can you do about this? Be a pack rat. Stockpile whatever you need based on who, what, where, when, how, and why forecasting. Or live on a self-sustaining farm. Where is your "castle"? Your home is your stronghold. Having spent decades in law enforcement, both as a patrol officer and a detective, in the Army and in Texas, I've worked more home and business burglaries than I can count. While many burglaries remain unsolved, I personally arrested every home invader I pursued. Some of these home invasion cases involved rape, attempted murder, and homicide.

Those stories, some of which appear in my *Dead Right There* detective memoirs-book, shaped my understanding of "castle crimes." Year after year, the numbers stay consistent. Where do they get in?

- 34% of burglars enter through the front door.
- 23% break in through first-floor windows.
- 22% use the back door. 9% slide under a slightly open garage door.
- 6% enter through other unlocked openings.
- 4% break in via the basement.
- 2% gain access through second-floor windows.

First, let's get the terminology straight, something that tends to irritate law enforcement professionals. Many people confuse terms like burglary, robbery, strong-arm robbery, armed robbery, and home invasion. Each has distinct legal and tactical meanings.

- Burglary: When your residence or business is broken into and no one is at home or work, it is

just a burglary NOT a robbery. It's a property crime.
- Robbery: A robbery is between people. A strong-arm robbery is when the robber has no weapon. He may threaten or mess you up. (If he messes you up seriously it might become aggravated robbery). Should the bad man or bad lady have a weapon, then it becomes armed robbery.
- Home invasion: A home invasion is when a criminal enters your home while you are there. The most dangerous.

Invest in a high-quality security system. A parked car in your driveway is major deterrent to burglars. At the risk of providing more of the crime prevention pamphlet-conundrum, can I please get you to read over this list? I found these public service lists from Strada Air Conditioners, Crime Prevention USA, common crime prevention pages and my own personal notes.
- Keep Your Home Locked Day and Night. To reinforce your locks, you can install a deadbolt protector. Many home intruders know how to pick locks, but a deadbolt protector prevents your deadbolt from turning so the door can't open. Reinforce your doors. In addition to keeping doors locked, you should reinforce them for additional security. If your doors have external hinges, you can reinforce them by replacing one of each hinge's screws with a security stud that locks hinge leaves together if a burglar removes the hinge screws.

Garage doors are another entry point to consider. You can secure a track-lifted garage door with a C-clamp by tightening the clamp on the door's track next to its rollers.

- Wood! Check the wood around your windows and doors to ensure it's rot-free is also helpful because rotting wood is easier to pry through. Hire a professional to replace any weakened wood around your home's entrances and windows with solid wood so it's sturdy enough to keep intruders out.
- Block sliding doors from opening. Whether they are locked or not, sliding patio doors can be easy to open. You can secure your home's sliding doors by placing a metal bar or pipe on your door's track. Cut the bar or pipe to the exact length of the track's space you need to fill, then place it there whenever the door is closed. You can also purchase an adjustable anti-lift lock for sliding doors. This type of lock attaches to your patio door's frame and the side of the sliding door panel to prevent break-ins.
- Glass houses! I am always paranoid for myself and all others because we live in glass houses where criminals can easily throw rocks. Replace glass doors. Another important factor related to glass doors is how easy they are to break. Breaking glass is an easy way for burglars to reach in and unlock a door, so you should consider replacing any glass panel doors or applying security film to the glass. Security film strengthens the glass and

prevents potential intruders from shattering it.
- Set a password for your Wi-Fi network. Protecting your Wi-Fi network may seem unrelated to home security, but it is one of the most important home security safety tips. Unprotected Wi-Fi networks are easy for hackers to access, and they can use this access to gather your personal information, monitor your house or control your wireless security system. Creating a strong password, keeping it private and changing it periodically keeps your security system and online information safe from hackers.
- Implement visual deterrents. Whether you have a security system in place or not, you should place home security signs in your yard. The visual indication of a security system helps discourage potential burglars because they typically target the least protected homes to avoid getting caught. You can also unnerve potential intruders by placing dog toys or a "beware of dog" sign in your yard. A dog could attack or make noise, and burglars try to avoid detection. Trimming the greenery around your home is another way to discourage intruders from approaching your property. The lower your trees and bushes are, the less burglars can shield-conceal themselves from your neighborhood's view.
- Remote control or times on and off lights confuse some burglars.

- Mark and inventory your valuables. In some situations, an individual's friend, neighbor or acquaintance are to blame for stolen property. While people often associate burglaries with unfamiliar intruders, some crimes are committed by individuals that homeowners know personally. Marking and cataloging your valuables makes it easier to retrieve them if a stranger or familiar acquaintance takes them. You can use an ultraviolet (UV) pen to mark your important possessions. A UV pen leaves an invisible mark that shows up under a UV or fluorescent light. When you mark an item with a UV pen, you can use the marking to prove an item is yours if someone takes possession of it or takes it to a pawn shop.
- Keep large purchases private. Large appliances and TVs or other valuable purchases can attract unwanted attention from potential burglars. People can see the boxes you place in your recycling bin, giving them a glimpse of the valuables your home contains. Hiding these boxes is an easy way to keep your home's contents private. Break your boxes down and fold them so the inner surface shows and the brand names remain unseen from the street.
- Address indoor safety concerns. Home intrusion is an important concern when protecting your home, but you should also consider how to protect your family from indoor hazards. Smoke and CO_2 detectors protect your family from catastrophic events

and illness, and medical alert buttons can help family members call for emergency response quickly when necessary.
- Install proper lighting. Lighting is an important part of preventing break-ins. Consider how you can make your home safer with the following lighting tips:
  - Install motion detector lights: Detector lighting can help deter burglars who may approach your property at night. A motion detector light turns on when it senses movement, which can send an intruder running if they think they've been seen.
  - A sudden bright light may make them think you're home or your neighbors can spot them. Place lights near doors: Many homeowners illuminate their front doors to deter intruders.
  - Install lighting at your back door and by any side doors your home has.
  - Installing lights by your garage, driveway and fence can increase your security and enhance your home's appearance.

Have a safe at home? On safes. How to secure a safe in your home: A safe may be one of the most crucial items in your home you need to protect, as it likely holds important documents, money or valuable items. A safe keeps your items locked away, but burglars often attempt to remove safes from houses.

You can increase your safe's security and prevent burglars from taking it or breaking into it with the following tips.

- o Bolt your safe to the floor or a wall. If there are pre-drilled holes in the back of your safe or in its base, you can bolt it to a wall or floor. You will need a power drill, a strong drill bit and anchor or expansion bolts to attach your safe effectively.
- o Keep a diversion safe in your home. In addition to hiding your safe, you can add a diversion safe to distract burglars. Purchase a small safe and use it to store lesser valuable items. You could place some cash or jewelry inside to make a burglar think they found your real safe, leading them to exit your home before they find your real valuables.

Are you safe at home? Many people feel annoyed when asked by a nurse or doctor, "Are you safe at home?" during their visit to a medical office. However, it is important to remain a patient, patient with this question because it is a small but crucial step in addressing the widespread issue of domestic violence.

Home safety is a significant concern. Although the concept of "the castle," one's home as a place of security was reported in an earlier chapter, the unfortunate reality is that many homes are sites of violence. A lot of domestic violence occurs at home.

Understanding domestic violence. Domestic violence can take many forms, including physical, sexual, emotional, economic, psychological, and technological abuse. It involves coercive behaviors that exert control over an intimate partner, such as intimidation, manipulation, humiliation, isolation, threats, and physical harm.

One of the strongest indicators of future domestic violence is strangulation. I attended a legal session in Texas focused on this specific crime. Studies have shown that non-fatal strangulation is a key predictor of future attempts or completed homicides. According to ResearchGate.com:

> "Non-fatal strangulation was reported in 10% of abuse cases, 45% of attempted homicides, and 43% of homicides."

Reiterated here for those reading out of sequence, individuals are often harmed or killed by someone they know, family, friends, or acquaintances. Intimate partner violence (IPV) affects millions worldwide, transcending race, age, and socioeconomic status.

Fortunately, most cities and states provide outreach programs to help domestic victims escape these dangerous situations.

My home invaded: A Case Study. I believe my home was once targeted by invaders, but I was able to thwart them with my firearm. Here's what happened and some key warning signs to help you protect yourself from this serious crime. Home invasions often go beyond theft, leading to violent crimes such as assault, rape, and even murder.

My Experience. At the time, I was retired and recovering from head surgery due to skin cancer. My head was stitched, bandaged, swollen, and I was weak from medication. During my recovery, a series of home invasions had been reported in our North Texas county. The pattern involved young women knocking on doors at night with fabricated stories, luring homeowners into opening their doors. Once the door was open, hidden male accomplices would rush in, leading to violent confrontations.

One cold winter night, while I was in a robe and pajamas, someone knocked on our front door. My wife, Jane, looked through the peephole and saw three young women in their late teens or early 20s. No males were visible.

"What?" I asked.

"Three girls," she replied, feeling inclined to open the door.

Before I could react, she reached for the lock. Instinctively, I grabbed my gun. As she partially opened the door, one of the young women pleaded,

"Help us. Our mom dropped us off at the wrong house across the street, and we're freezing. Can we come in and call her to come back?"

Jane, concerned for them, hesitated before saying, "Wait just a minute." She closed the door almost completely and turned to look at me. By then, I had my gun aimed at the door and shook my head, signaling "no."

Still, Jane felt sorry for them and reopened the door slightly, she said later to offer to call their mother and/or give them blankets to wait outside. At that moment, the first girl barged in, and right behind her, a young man followed!

Without hesitation, I pointed my gun directly at him and commanded, "Get out or die."

He froze. He had no visible weapon. If he had pulled a knife, a club, or a gun, I would have shot him immediately. Jane wisely stepped back. Both the intruder and the young woman backed up and left without a word. Jane quickly shut and locked the door.

"Call the cops," I instructed her, which she did.

When the police arrived, they took our report. I remember how they eyed me up, bandaged, swollen, and in no condition to physically fight or detain anyone. I wanted them to be aware of these suspicious individuals in our neighborhood. However, the group was never found.

Were they simply lost, as they claimed? Perhaps. But given the local crime pattern, their behavior was highly suspicious. They disappeared so quickly, reinforcing my belief that this was an attempted home invasion.

Ways of home invasions – please take notes to prepare:
- One big safety tip? Just don't answer the door. Especially after hours.
- THE SNEAK IN. The suspects secretly inspect your house and enter where they can while you

are home. These criminals don't always ring the doorbell, and they enter in another way and surprise you.

- THE FRONT DOOR RUSE. Like the previous story. They might be strangers to you with a big smile. Lost? Survey? Salesperson? They might have known you or know someone that knows you from the present or past. One of the home invasion murders I worked involved a suspect that knew the victims. He'd worked on their roof in the past. Then he returned, beat and robbed them, and left the old couple for dead.

  One case I worked involved two Las Vegas mobsters that knew "from the grapevine" that the victim had gold bullion at home. Masked, they knocked on the door. Knock answered? They charged in.

- THEY FOLLOW YOU HOME. They might follow you home from, for example, an expensive restaurant (several of my cases). As you pull in your garage and before you close the garage door, they barge in, weapons displayed.

I know anti-gun people like to proclaim that you shouldn't have a gun handy at home because in some very rare occasions, presumed home invaders are legit relatives or friends, etc.

Here's a tip, evaluate! And don't shoot them! Then naysayers complain about handy guns on very rare occasions can lead to handy suicides. Then handy guns lead to other very rare accidents. All tragedies, yeah. But in the big population picture, like in the United

States where there are 340-plus million people, with millions of more guns in comparison, these tragic events are still extreme, tragic rarities.

As I warned, home invaders usually have more on their minds than just simple theft. Usually rape, aggravated assault, torture and murder. I still feel sorry for people in states and countries where they cannot have a gun in their castle to combat these home crime and times. When you need one? You really, really need one.

Home invader in your house? Usually at night. Disagree or not, the generic public safety advice is get your gun, get down beside your bed, call the police. Describe everything happening. Great for the solo person. But what if you have kids in other rooms? It would be hard to follow these hide-away, ever-so-generic commands. I haven't and I won't advise you, but I have probably searched a thousand homes and businesses through the decades, gun at the ready, why stop now? But, that's me. You will have to make that "hide or search" decision for yourself.

Final thoughts on home invasions. Home invasions are a terrifying reality. Always stay alert and trust your instincts. Never open your door to strangers, especially at night. If something feels off, call the police immediately.

Leaving your home for temporary "homes." Where, when you are away from your home, could you have problems? On the move away your castle. Whether a professional or a civilian, where can your routes and daily stops of your life be dangerous? List where you go every day, every week. Month? Year. Where are you staying. Where-how do you get back? From the mailbox out in front of your house, from the supermarket, from the job, from the vacation? Then list what could probably happen on these small and big trips. Then make a list of what could possibly happen.

Crisis rehearse. Where are you? When way? Vacation? Business trip? Plan with a *Safety In Mind* report by Berkshire-Hathaway. BH published a very comprehensive list I think you should read over.

1. Allow plenty of time to do safety focused research. Give yourself ample time to plan for your trip. In any case, starting your planning early gives you more time to research where to go and the specific sites to see. And while you're doing that, you can make safety part of your research criteria.

If you're sticking to the U.S. you have plenty of resources to inquire about safety concerns and travel advisories. Check out the websites of state and local governments and tourism boards, and do some smart web searches like, "Is it safe to travel to ...?"

2: Secure the homefront before you go. Before you leave, make sure you secure your home and take basic precautions like the following. Note that all may not apply to you and your home:

- Stop your mail, newspapers, and other deliveries. You don't want to broadcast that you're gone. Or, get a friend to collect your mail.
- Let a trusted neighbor know what you're doing and ask them to keep an

eye on your place. If a package does happen to get delivered, they can also pick it up. Secure valuables in a hidden safe or off site, like a safety deposit box at your bank.
o Lock doors and windows of every kind on every story/floor. Also, stash your extra keys away, especially the ones you've "hidden" outside your house.
o Double-check that your alarm system is good to go.
o Confirm your security lighting, motion detection lighting is working.
o Make sure your closed-circuit television (CCTV) is functioning, has the capacity to store footage, and will work if you plan to access it via the internet while you're gone.
o Schedule lawn maintenance. Overgrown plants and long grass can make it seem like your home is vacant. Burglars and vagrants sometimes target homes with unkempt lawns so they can commit theft, or stay, with a lower chance of getting caught. A well-maintained lawn indicates that someone is likely home.
o Ask a friend to stay in your home or use your driveway. Again, a car in the driveway is an *INSTANT* burglar or vagrant turn-off.
o Be sure your cellphone will work where you are traveling.

- Alert your bank and credit card companies. Of course, you want access to funds for the day-to-day activities of your vacation. Drawing from an ATM thousands of miles away could look like suspicious activity on your bank account.
- Crime statistics prove that visiting places like hotels, motels, restaurants, gas stations on or very near highway (aren't most of them?) can be inherently more dangerous. FBI interviews of many convicts reveal that they wanted the fast getaways that highways allow for, versus police response times.
- Don't dress like a tourist. Don't dress like a paid contractor unless you have to for a mission advantage. You might become a target.
- Someone you know and trust should know your itinerary.

Be smart about where you are going. Years ago, I wanted to fulfill my wife's dream of seeing Paris, France. I arranged this as part of my teaching schedule. She began investigating Paris and read about 6 common scams that the gypsies pull. She handed me the list and I read it. When we toured Paris we were hit by all 6 of these schemes.

A very thorough and fantastic book about the small and big details of travel, which include safety tips and advice of world travel is *Choose Adventure* by my retired cop, friend, Greg Ellifritz. Greg is addicted to traveling to obscure places and documenting every iota of the trip, but also with a cop's eye. This is no ordinary travel book. Not at all. It is a very unique deep informative and educational dive about travel you will not find anywhere else.

WHERE Question 2: Where do you work and how dangerous is it there? How safe are you at your workplace? In the United States, the Department of Justice and the Center for Disease Control try to track injuries and acts of violence in the so-called sweeping term "American Workplace" that arena where in a nation of some 340-plus million people, of whom teen-agers and adults toil at some job.

Other civilized countries have their own official groups keeping track of this issue. Excluding the vast amounts of workplace accidents and injuries, any place were people interact is a natural touchstone for interpersonal violence.

Sometimes the simple difference between these categories is good first aid and great life-saving EMTS and ERs. I give you the Industrial Safety and Hygiene News recent compilation for U.S. Bureau of Labor Statistics BLS). Remember this list will change from year to year, (thus, the "Often Number," nomenclature) but generally remains the same. Do peruse the list because I think some categories may surprise you. For those reading this outside the United States, these dangerous jobs are something to consider within the context of your country too.

The job danger list from the U.S. Bureau of Labor Statistics: In pursuit of seeking that "big picture" idea I introduced earlier in this book, here is a big picture look at dangerous jobs. This is drawn from general annual averages. I think you will find some ratings a surprise. You might be among them.

- Often Number 1. Logging workers.
- Often Number 2. Aircraft pilots, flight engineers.
- Often Number 3. Derrick operators in mining.
- Often Number 4. Roofers.
- Often Number 5. Garbage collectors.
- Often Number 6. Ironworkers.
- Often Number 7: Delivery drivers.
- Often Number 8. Farmers.
- Often Number 9. Firefighting supervisors.
- Often Number 10. Electric Power linemen.
- Often Number 11. Agricultural workers.
- Often Number 12. Crossing guards.
- Often Number 13. Crane operators.
- Often Number 14. Construction helpers.

- Often Number 15. Landscaping supervisors.
- Often Number 16. Highway maintenance workers
- Often Number 17. Cement masons.
- Often Number 18. Small engine mechanics.
- Often Number 19. Supervisors of mechanics.
  Often Number 20. Heavy vehicle mechanics.
- Often Number 21. Grounds maintenance workers.
- Often Number 22. Police officers. Police and sheriff's patrol officers. How dangerous is it to be a police officer? Working as a police officer is about 4.1 times more dangerous compared with the average job nationwide, based upon the workplace fatality rate. Police officers have a workplace fatality rate similar to maintenance workers, construction workers, and heavy vehicle mechanics. The most common causes of death for police officers at work are traffic accidents and violence by persons. And something else to consider about police work...

    "The average person experiences two to three critical incidents in their lifetime, while police officers can experience up to 178 critical incidents during their career. A critical incident is defined as an unscheduled event that involves potential injury or property damage and requires a law enforcement response. Examples of critical incidents include:
    * Officer-involved shootings
    * Child abuse
    * Vehicle accidents

* Death or serious injury of a co-worker
* Line-of-duty death
* Gruesome homicides
* Police officers are often exposed to repeated crisis situations. Research has found that peace officers experience an average of around 3,45 traumatic events for each six months they serve." - The Impact of Life Experiences in Police Officers, By Saul Jaeger, M.S.LEB, FBI.gov

- Often Number 23. Maintenance workers.
- Often Number 24. Construction workers.
- Often Number 25. Mining machine operators.

Often, we forget other dangerous jobs, such as in the medical field, where the U.S. Center for Disease Control lists "problem people" as a source of trouble and violence. I report this list because they are rather common, universal "people problems" everywhere. Citizens, soldiers and police will meet these problem people in everyday life, before or after they are patients. This list of dangerous people includes those who:
- are under the influence of drugs or alcohol.
- are in pain.
- have a history of violence.
- have cognitive impairment.

- are in the forensic (criminal justice) system.
- are angry about clinical relationships, e.g.
- are angry in response to perceived authoritarian attitude or excessive force used by the health provider have certain psychiatric diagnoses and/or medical diagnoses.
- It is important to realize that, although some psychiatric diagnoses are associated with violent behavior, most people who are violent are not mentally ill, and most people who are mentally ill are not violent. Substance abuse is a major contributor to violence in populations both with and without psychiatric diagnoses.

Workplace violence. In a given year, there are hundreds of homicides recorded in workplaces. According to the Department of Justice, on average, 1 in 6 violent crimes occur in a workplace. Some 2 million American workers are victims of workplace violence each year.

Murder is the leading cause of death for women on the job. It seems these days like each week we see or read in the news about another man showing up at his wife, or his ex-wife's or his girlfriend's workplace and shooting or beating her. Often he injures someone else in the process. Often this male then kills himself. A common outcome. Where are the common workplace violence danger zones? What is workplace violence? The U.S. Occupational Safety and Health Administration (OSHA) definitions and advice are worthy worldwide.

> "Workplace violence is violence or the threat of violence against workers inside or outside homes and buildings, ranging from

> serious threats/verbal abuse to physical assaults and homicide. However, it manifests itself, workplace violence is a growing concern for employers and employees nationwide."

Workplace violence can strike anywhere, and no one is immune. Some workers, however, are at increased risk. Among them are workers who exchange money with the public; deliver passengers, goods, or services; or work alone or in small groups, during late night or early morning hours, in high-crime areas, or in community settings and homes where they have extensive contact with the public.

How to best create a safer workplace is a HOW question seen later in this book.

On the subject of business robbery. There are some broad tips that range big. Robbery victims are both customers and employees and cross the lines of many people and businesses. Whether working or shopping or walking, watch customers for suspicious activity and clothes. Robbers often "case" the premises. They often eye the layout of the business and other customers above and beyond what normal customer, product seeking and transactions. They often wear clothing and hats that can partially conceal or fully conceal their faces. They often look for video cameras. They often wear clothing that is contrary to the weather to conceal weapons.

What are the bank employees told and taught? This advice is handy for all tellers and counter clerks, and to some extent, citizens within the robbery area, making the advice really worth reviewing.

In the United States, The Bank Protection Act requires that all employees and officers be trained annually on proper procedures for robberies, larcenies and burglaries. Most financial institutions will train their tellers but fail to recognize the importance of training all employees and officers regarding their responsibilities under the protection act. Other countries have a similar law to follow.

Tony Brissette is a veteran of over 30 years as a Director of Security in Massachusetts. Now President of Brissette Consulting Services, Inc., in Shrewsbury, MA, he specializes in bank security training programs. Tony advises (and think of how these tips relate to all store employees and customers.):

"In my capacity as Corporate Security Training Officer for a major bank located in Boston, Massachusetts I received reports on all bank robberies that occurred in the New England region. Most of these robberies were handled well by tellers and employees in the proper manner but from time to time an employee would either not comply with the robber's demands or would do something that escalated the level of danger and exposed both employees and customers to increased danger.

The following list of ten catastrophic mistakes committed by bank employees and customers based on those reports: (also take note of how many of these tips relate to you being present during and business robbery, not just banks,

    1- Do not treat the holdup note as a joke or a prank. There have been several instances in which a teller has been handed a hold-up note and believes the customer is joking. If the teller does not believe the note is serious, the robber may feel forced to display a weapon, escalating the likelihood of harm.

2- Do not create any surprises for the robber. In some cases, tellers have walked away from their teller station if they don't observe a weapon. Others have been advised to pretend to faint. These actions may be successful in thwarting the bank robber, who may simply run out of the bank. But if the robber is really desperate, the teller's actions may cause the robber to display a weapon and possibly grab a customer in the lobby. Do exactly what the robber tells you to.

3- Do not carry excess cash in your cash drawer. Bank robbers will come back if they're given large amounts of cash. Tellers should adhere to their bank's cash limits for both top drawer and teller station. If a teller accepts a large cash deposit, excess cash should be transferred to the head teller immediately.

4- Only give the robber the money demanded. Don't ask if the robber wants the cash in your second drawer.

5- Do not attempt to bring attention to the robbery. Statistically bank employees who follow the bank robber's instructions are seldom injured in the course of a robbery. Handle the bank robber as you would a regular customer. Don't attempt to gain the attention of anyone else to alert them to what is going on. The most important role you have in this robbery is to ensure the safety of all employees and customers in the bank.

Bringing attention to the robber could compromise the safety of all.

6   Do not argue with the robber or attempt to talk him/her out of the robbery. Arguing, confronting or attempting to talk the robber out of the crime will increase the likelihood that others will become aware a robbery is in progress and escalate the level of danger.

7   Do not tell customers that you have just been robbed. After one robbery, just as the robber reached the front door, the teller yelled out "grab him, he just robbed me." This was an extremely dangerous action that places the safety of employees and customers in danger. What if a customer did attempt to grab the robber and a struggle took place in which a weapon was used and either a customer or employee was injured or killed? On occasion customers, believing they are acting as good Samaritans, have confronted or chased robbers and increased the likelihood of danger to themselves and others.

8   Do not ever leave the bank after a robbery. In numerous cases we found that after the bank robber left the branch, a bank employee will either exit the bank to see if they can observe the robber's getaway or - worse -actually pursue the bank robber in a chase. This type of action not only places the employee in danger but also poses a threat to others. If during such a chase

or attempt to observe a robber's escape and someone is injured, the bank will have potential liability in a possible civil action. This is an especially important message for non-retail employees who may not have been trained properly. Let the police chase the robber.

9- In a takeover robbery, do not make sudden movements. Do not attempt to activate hold-up alarms, run out of the bank, or attempt to call the police. Takeover robberies are extremely dangerous because the robbers are most likely displaying weapons. If you are on the telephone when a take-over robbery occurs hang the telephone up and do not answer any incoming calls unless the robbers tell you to. Attempting to activate an alarm can also be very dangerous if the robbers observe you during your attempt. Do not try to escape the robbery, as robbers will be closely watching for this activity.

10- Do not ever attempt to engage the robber(s) in a struggle. Although most of us would never imagine engaging a robber in a physical confrontation, there have been cases in which bank security guards, branch managers and other employees have physically confronted bank robbers. Remember this type of response to a robber increases the level of danger to all employees and customers in the bank.

11- *My Note for this:* Decades ago, owners, managers of businesses, certainly eateries, bars, etc. ended their day with lots of cash. Some had safes.

Some took that day's take to a night deposit

drop at their closed, off-hours bank. End of the day is a juicy target for armed and/or strong armed robbers. They rob the establishment at night or...rob the person dropping the night deposit off. Think about all the banks you know. Think about how secluded and dark some night drop-off, locations are. This was always a problem. We told drop-off people to circle the bank once or twice and make a really good safety study of the area before parking and dropping the money bags off. Go through the Ws and H for such a safety check.

I don't think much of these night bank deposit, drop-offs occur in our modern times, but if it does, if you have to do this, take heed of this old school advice of doing a recon around the bank before parking.

What are workers and customers taught and told by authorities? Every law enforcement official will tell you the same things. You should not try to stop the robbery by force. Keep in mind the FBI reported that an increasing number of robbers are using powerful drugs (cocaine and methamphetamine) during the time of the robbery. You don't want to take chances with a dull-thinking, or tanked-up person in such a desperate situation. If you notice that robbery is in progress, simply get some details on the robber that you can pass on to the police. Some helpful information:
- o Note clothing. And look for layers under the visible layer (the outer layer can come off as part of the escape plan).

- Note physical characteristics – height, weight, eyes, hair, mannerisms, scars, etc.
- Note automobile description (not the most important since it's probably stolen).
- Note last known direction of travel.
- Note weapons – the police need to know if the robber is armed, and with what?
- Note...anything else! Try to remember everything.
- Note to not disturb any evidence at the scene.
- The experts say it's best not to attract any attention to yourself. Simply follow instructions so that the robber can get out of there as soon as possible. Law enforcement will pursue the robber.
- Agitating a robber has statistically resulted in harm to yourself and others.

Where home and workplace violence meet. I will say with confidence from my training, experiences and study that a lot of violence occurs anywhere over domestic relationships like cheating, money problems, and divorce. Incidents of violence involving frustrated relationships and family violence. These events are often carried out over to the workplace. Angry spouses show up at work with guns. People get shot. A worker should be aware of, even if from gossip, that romantic and domestic turmoil often extends into the workplace. Making your workplace safer is coming up in a HOW segment.

WHERE Question 5: Where can you sit, stand, run, hide or fight? When you enter homes or public buildings, schools, stores, restaurants, churches, etc., take a moment to assess your surroundings, "Jason Bourne" style (if you're old enough to remember those movies). Use sensemaking (as defined earlier) to evaluate who is present, what's happening, where you and others are, and where you're headed.

Is there any sign of trouble? Why are you there? Why are you still there? What potential threats, criminal or terrorist might arise? And if a problem does occur, how will you respond?

You've probably heard the advice to sit with your back to the wall in public spaces for better visibility. While that's not always possible, consider it.

If a situation becomes dangerous, how many ways can you get out, and quickly? Will those exits become congested? If a robbery happens at the register, how far are you from the specific danger?

Think about safety under gunfire. In military and law enforcement, the terms "cover" and "concealment" are key. Cover stops or significantly slows bullets.

If near cars and outdoors, experts recommend using a car's wheels and engine block for protection. Indoors, assess the density of furniture and objects, look for the best available cover.

Concealment, on the other hand, won't stop bullets but might keep you hidden when necessary. Sometimes, staying out of sight is the only best option.

WHERE Question 6: Are parking lots like those at Walmart the most dangerous places on the planet? A special "big picture" study. This topic deserves its own in-depth discussion, as it ties into the broader examinations we introduced in an earlier chapter. Parking lots warrant a deeper analysis and their own category of inquiry.

It's important to note that crime statistics collected by the FBI are often described as "patchy" by experts. Despite this, retail stores with parking lots frequently make headlines due to safety concerns. But how significant is the risk?

According to the National Institute of Justice, nearly 10 percent of all violent crimes in the U.S. occur in parking lots, whether at businesses, schools, churches, apartments, or other locations. To put this in perspective, CNN estimates there are approximately 2 billion parking lots in the United States. If 10 percent of violent crimes take place in parking lots, this amounts to roughly 130,000 incidents per year.

That sounds like an alarming number. Media reports often fail to provide this broader context. In 2023, Business Insider reported that employees at Walmart and Target worry about safety inside the store and on their parking lots, not only due to violent crimes such as muggings, assaults, and homicides but also because of reckless driving and parking-related altercations. Just as road rage is a concern, so is "parking rage.

"Walmart, in particular, is frequently scrutinized in the media, often portrayed with images of unusually dressed customers and accused of negatively impacting small businesses. If you follow the news, you might assume that crime is rampant in Walmart parking lots. A

This Week magazine headline once proclaimed, "Criminals Flock to Walmarts," yet the article contained no statistical evidence to support this claim.

For the purpose of analysis, I will use Walmart as a case study in parking lot crime. My goal is not to defend Walmart (or any parking lot) but to examine the data within the broader context, a perspective often overlooked, whether due to ignorance or bias.

- Walmart numbers 1: As of July 31, 2024, Walmart had 5,205 retail units in the United States.
- Walmart numbers 2: Walmart's net sales for 2024 are estimated to be $642.64 billion. This is a significant increase from previous years.
- Walmart numbers 3: According to the most recent annual report, the average Walmart location serves approximately 35,000 customers per day. Times 365 days a year, based on that, there are about 12,775,000 transactions.
- Walmart numbers 4: Based on the layouts and locations of Walmarts, one could easily assume that almost all of the transactions include one car parking on a Walmart parking lot per sale. In the broadest "patchy" terms, that means about 12,775,000 cars parking on their lots, with one or more people coming and going.
- Walmart numbers 5: After my examination into Walmart crime, it is impossible to determine how many crimes were on the parking lot juxtaposed with crimes like theft, inside the store.

- USA violent crime numbers. Based on data reported to the FBI's Uniform Crime Reporting (UCR) program, which covers a large percentage of the U.S. population, the estimated number of violent crimes for 2023 was approximately 1,278,301. (This book is written in 2026), The Department of Justice-Federal Bureau of Investigation are perennially, NOTORIOUSLY behind on these number collections and crunching.
- The NIJ estimates an annual 10% of the national number of violent crimes occur on parking lots they would guess-estimate 120,000 parking lot crimes.

Possible patchy summary: Let's use the lowest number, since we cannot retrieve exact Walmart parking lot crime numbers. Even if all 120,000 violent store crimes occurred only on Walmart parking lots, that would be 120,000 crimes connected with some 13 million cars of one or more people. Even with this worst case, less than 1% of all customers.

That's pretending all parking lot crimes in the country occurred on Walmart lots. If we had the real numbers, per person crime rates on Walmart lots would still be miniscule. We might even add that violent crime on any parking lot in comparison with total cars and customers, on average would be minuscule.

But study your local problems. Renown gun instructor, ex-cop Tom Givens was headquartered for years in Memphis, Tennessee. Many of his gun course graduates, 60-plus of them, were survivors of Memphis area, parking lot gun fights. But Memphis always has a higher violent crime and property crime rate than most other cities in the United State and the world. Tom has saved a lot of Tennessee lives with his instruction, and not just there but around the country with his precision teachings.

Where did you park and car theft and car burglaries. That's enough of violent crime on lots. Now what of property crime like car theft and burglaries? Take a deep breath, grab some more of that coffee as we list these common security tips.

- Lock it and pocket the key whether you leave for a minute or several hours. Close windows all the way and make sure the trunk is locked. (Unless you live in a liberal extremist city these days where you are told to leave your windows down and leave nothing of value inside the car, else criminals will break your windows for a car burglary.
- Control your keys. Never leave an identification tag on your key ring. If your keys are lost or stolen, this could help a thief locate your car and burglarize your home.
- Don't leave your registration inside your vehicle but carry it with you. Important identification papers or credit cards should never be left in a glove compartment.
- Keep everything of value covered in your or in the trunk. If you do leave packages, clothing, or other articles in the car, make sure they're out of sight or covered.
- Park in well-lit and busy areas. Avoid leaving your vehicle in unattended parking lots for long

periods of time. If you park in a lot where you must leave the key such as with a valet, leave the ignition key only. (The whole key ring with your house key on it could become a house burglary, home invasion problem later.)

o When buying a car, ask about anti-theft options such as steering column locks, alarms, switches that interrupt the fuel or electronic systems, and locks for batteries, and gas tanks. Many insurance companies offer reduced rates to owners who install security devices.

o Keep your car's VIN (Vehicle Identification Number) and a complete description in a safe place at home. Since 1969, the federal government has required manufacturers to engrave a unique number, the VIN, on all passenger cars in one visible and several hidden locations. One VIN is engraved on a metal plate on the dashboard near the windshield. VINs are registered with the FBI's computer National Crime Information Center.

o License plates frequently are stolen from other cars and used on stolen cars for other crimes. Get in the habit of checking your plates when you drive. A few drops of solder on the bolts or blurring the threads will help safeguard your plates from easy removal.

o With your installed car alarm - a car alarm can help prevent vehicle theft. (Often, safety experts suggest that if you keep your car keys nearby, especially at night, and you and the keys are near your garaged car or otherwise

parked car, you can hit the emergency alarm button on your fob, causing some level of attention to your location.)
- Operators who work in volatile locations (like embassies and so forth, or contractors in controversial locations like factories, etc. worry about riots. Sometimes you can park away from common-sense-projected riot and demonstration areas. Can you get to your car though? Depends on your Ws and H projections. In really hot zones you have to worry what opposition might do, covertly and overtly to your car.

    I have taught in various Southeast Asia (Middle East) bases and to and fro vehicles go under such a heavy inspection that you cannot watch what inspectors do. You are sequestered in a no-see location while they super search.

Parking lots are indeed open invitations to meet and pass by all sorts of humanity. Just because we can statistically refute stories of presumed high, sometimes "war-zone-like" numbers on lots, doesn't mean you shouldn't take a good look around you when on one and use the Ws and H analysis to the fullest extent.

Be aware of where you park and if people around you look and act as suspicious as those in the prior lists of this book. Hide any valuables inside before you leave the car. Scan the area around your car as you approach. If you see someone loitering around your vehicle, walk past until they leave. Consider using a shopping cart rather than carrying bags on and, or in your arms. This way you are free to maneuver around the cart or escape.

You can let loose of cart the handle and fight or draw weapons.

While it is possible to use a shopping cart for protection, it's a makeshift, emergency, self-defense tool, and the top priority in a dangerous situation should always be escaping to a safer area. The shopping cart can serve as a buffer, obstacle, or improvised weapon to create distance and disorient an attacker.

When onloading and loading, you can open a side door and pull the cart up behind you, like in a triangle, for unloading ease and protection.

Criminals are also known to use "windshield bait." That is sticking something momentarily distracting on your windshield or your door window. Thet are nearby waiting for you to be curious, look and read the sheet of paper or whatever the bait is.

My "Parking Lot Madness drill.

This photo on the next page is from my *"Shooting In, Out and Around Cars"* Gun Module. I call this drill "Parking Lot Madness." We set up a parking lot of seminar attendee cars and using electric pellet guns (they don't hurt cars, gas guns on up, will damage cars), I run one-on-one (or more shooters) gun fights.

We have used electric pistols and even electric machine guns. Handgun-wise this event could happen to anyone. Machine gun-wise, it's a great exercise for the military and SWAT.

WHERE Question 7: Where and how am I going to stand in a confrontation? Through the chit-chat, the interview, the argument, the fight, and the surprise ambush possibly within each, you might stand...

- The Bus stop "day dreamer" - unaware. Inspired by Kenpo Karate's Ed Parker, this is when someone is standing oblivious to a pending problem and must "spark and leap" mentally and physically into action. This is the one "un-ready" position amongst the ready. (Of course, you can daydream while seated too.)
- The Bus stop stand again, yet aware. This is the set-up for "sucker-punch" action.
- Ready stance right lead, in something of a right shoulder, right leg forward lead. Your choice from training.
- Ready stance left lead, in something of a left shoulder forward-left leg forward lead, your choice from training.
- Knee-high fighting angled upward, versus standing opponents. (We couple this response material with a seated person too.)
- Knee-high fighting another knee-high opponent. (Have both fallen? Many times, a thrower drops you and he remains knee-high.)
- Knee-high fighting versus someone under you. (This is the top side of a hand, stick, knife or gun, floor-ground fight.)
- Down on your back fighting those above, grounded through standing.
- Down on your right-side fighting those to your right, grounded through standing.
- Down on your left-side fighting those to your left, grounded through standing.
- Note: There are suggestions for specific unarmed and armed positioning.

All of these are important, situational problems that cover the full spectrum of a real fight. Whatever you practice standing you should experiment with these other "heights" also.

The latest police training suggests that officers should stand "two giant steps and an arms lunge" away from a confronter- suspect. This is not a bad idea for the military when interviewing possible hostiles. And not a bad one for civilians either. But civilians might ponder, if things are heating up, why are you still there anyway? And do keep in mind even this 2 ½ distance can be covered very quickly.

I do not obsess on the term "fighting stance." A fighting stance is a still photograph while a fight is a 3-D moving motion picture with balance and power in motion. As my first JKD instructor in the late 1980s, Paul Vunak used to say, "There is no football scoring stance. There is no basketball ball scoring stance. There is no one fighting stance." There is just balance and power in motion. Obsessing about fighting stances is a time-waster, a distraction. Instead, we work around this ready list that preps people for the...3-D.

Some pre-fight, interview postures. Good idea to have the hands ubiquitously, nonaggressive, and higher up to take action. All strikes and kicks should be practiced from your selected postures. All this is found in my "how-to" books.

WHERE QUESTION 8: Where do the fight "collisions occur" the crashes, the sticking-stopping points occur in an argument and a fight? Some people refer to this as "ranges."

How many ranges are there in an interview, an argument and a fight? I would rather not use the term "ranges," but many citizens do ask this range-like question and numerous martial arts do officially designate ranges, like kicking range or boxing range, etc.

Instead, I like to use collision-crash-stopping points and I developed the *Collision 6* program. The program is not a course but rather a laboratory to create option lists and organize mixed weapon workouts.

There are 6 collisions, or the six common crash points in a typical fight (or arrest). I think this is important for the curious, the student, the practitioner. Briefly, here they are. You need my other books and seminars for specifics.

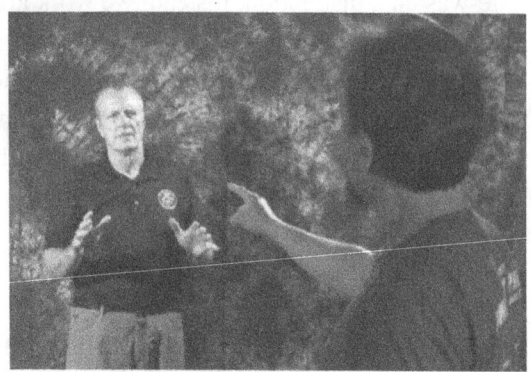

Collision 1: Collision of the minds. No physical contact yet. This starts out as far away as sniper range on up to a face to contact confront-argue distance. The common stand-off.

- *1) Alertness-awareness.*
- *2) Environmental scanning.*
- *3) Sensemaking.*
- *4) Verbal skills.*
- *5) Evasion skills.*

Collision 2: The Tangler! The common hand grabs. Fingers on fingers. Hands on hands. Hands on wrists. Hands on weapons from their weapon carry site to use. Pulls, grabs, counters to grabs and related scenarios.

Collision 3: The Forearm Collision. Forearmed is Forewarned! The common forearm-to-forearm collision. With and without, and against weapon pulls and weapon use.

Collision 4: The Arms Race! The common biceps-shoulder-to-neck-line collisions. After some stunning blows, this covers a lot of arm grappling, and takedowns. The arms are in range to hit, wrestle with capture, and even break.

Collision 5: The Bear Hug Collisions! (All arm wraps). The common bear hugs of the torso and legs (tackles) and/or clinch collisions.

Collision 6: Ground Zero! The common ground fight collision.

The less a person is trained, the worse one will do when "colliding." This is geared for any two-person encounter but could include more people. Even if you are inside a large hostile group or a riot, slivers of that group often fight one-on-one, or one versus two within. These stops are not classic martial art ranges but instead are verbal and physical collisions that take place in interviews, arguments, crime and war.

Obviously, the six collisions do not always occur in any particular order. Collision 1 could be last or Collision 6 the first. They are organized here this way to offer a training progression, a laboratory to train and experiment in. (All actual fighting methods within these six collision-stops-crashes are in other photo-laden, how-to books I've written and videos.)

WHERE Question 9: Where on the body should you shoot, hit, stab or slash an enemy? I often arrived at terrible scenes before EMTs and I had to apply what little I knew as fast as I could. I still include "gutter medicine" emergency medicine in all my courses. I even did a 1990s video on the subject with an EMT, years before the movement became popular with shooters and martial artists.

It would be wonderful if we all could attend an EMT academy, wouldn't it? But we can't. Still, we should know some basics on how to treat ourselves and others when hit, shot, stabbed or victims of accidents and disasters.

Let's start after the fact and work backwards. Reverse engineering. The most life-threatening? Knowing who, what, when, where how and why to treat those sites first, second and third, helps you to know the opposite where to attack. Duality again. It's called

Triage. When triaging, where on the body should you look and treat first on your wounded family member, comrade or stranger?

Ed Sizemore, a lead instructor at the Federal Law Enforcement Training Center's Firearms Division, tells his law-enforcement students to orient their bodies toward the threat. Most people doing the shooting tend to aim for the easiest target - the torso. Since police officers wear body armor, they have the most coverage in front and back. He would not, Sizemore makes a point to add, recommend the same thing to any unvested civilians. Turning sideways is less of a target, unless, of course, the target is very obese.

In Sizemore's opinion, there's no single best place to be shot. Ballistics, the study of the projectiles, like bullets is too much of a gamble. He says, "People get shot in fatal areas and live, and others get shot in non-lethal areas and die."

But he believes the most painful place to be shot would be in your pelvis. The nerve bundle located there would quickly and efficiently distribute pain throughout your body. He can also think of a worse place, medically speaking: "The heart can be repaired. There is such a thing as an artificial heart. But as far as I know there are no artificial brains."

I have worked numerous shootings through the years, and I can attest to the fact that the flight of the bullet inside the body is quite fickle. To save time I will give you one example.

One afternoon we detectives were chasing-hunting three robbers and one of us spotted them on a rural road. Our friend called the spotting in on the radio and got out of his sedan to approach them. One criminal was an ex-con, armed with a .32 caliber pistol. He pulled that .32 out and our detective began to draw his pistol. But action beat reaction. The first .32 round struck our man in the weapon limb forearm. The detective dropped his almost drawn gun. That same bullet went through the arm, hit his rib cage and circled around the front of the body and exited on the left side, whereupon that same bullet lodged into his left forearm. The first bullet! And our detective was disarmed with multiple injuries. The shooter emptied his gun into the now collapsed detective. Anyway, we got there and we arrested the trio after a long rural search. The detective lived.

The flight path of a bullet in the human body can be fickle! When you check for exit wounds, the exit might not be where you think it should be.

I can advise you from 40 years of Filipino stick fighting, earning two Filipino martial arts black belts, and a graduate of numerous old-school police baton courses, that stick strikes to the head and kneecap are devastating. But as with bullets, impact weapon blows to the elbow and other body parts can certainly be bad too.

Anywhere you are stabbed and slashed can be a serious problem. Muscles and veins are cut. Stabs to the diaphragm and heart, to the throat, in the eye! All of this is bad news. If they are not 'show-stoppers" you will have to continue to fight an assailant until their blood loss-pressure drops and they pass out.

WHERE Question 10: Where is the planned, second location? Look up these words on the internet - "Never be taken from 'Crime Scene A' to 'Crime Scene B.'" and all you will get on this subject is from "CSI Investigator Google' preaching about crime scene searching and evidence contamination:

"Do not be taken from crime scene A to crime scene B," means a body or piece of evidence should not be moved from crime scene A to crime scene B because doing so could potentially contaminate both scenes and compromise the integrity of the investigation by transferring evidence between locations."

Is that all there is about A moves to B? Is this all the interweb can pull up these days on this important subject. Reporting on evidence contamination. Even with the popularity of true crime serial killer shows on cable TV, as well as fictional shows? Contamination is not what I want to explore here.

What about life-saving information. I am talking about people abduction and the first and second location problem. The criminal tricks and/or controls, captures (crime scene A) and hauls the victim off to his favorite, obscure, torture place (crime scene B). A place where Hannibal Lector turns your brains into beans.

Should you get a chance to fight back, and it looks like a "takeaway"-kidnapping in progress at first contact – Crime Scene A? Fight, give it all you got at Crime Scene A! You do not want to be delivered to Crime Scene B.

For example: Theodore "Ted" Bundy was a serial killer who kidnapped, raped, and murdered dozens of young women and girls during the 1970s. He would then lure his victim with tricks (Crime Scene A) to a vehicle parked in a more secluded area, at which point he would try to bludgeon her unconscious, then restrain her with handcuffs before driving his victim to a remote location (Crime Scene B) to be sexually assaulted and killed.

Note: The ATM conundrum. While many robbery victims are confronted (Crime Scene A) and ordered to go to an ATM machine (Crime Scene B) many are then let go once they have stolen money. Yes, some are not set free, so if you can? The time to resist and fight back is crime scene A.

WHERE Question 11: Proprioception. Where is my body part now? I've lost track! No, not after you have been dismembered. Not at all. You are alive! Seeing and feeling. We see the world from our two eyes in the upper part of our face. When we can't see the rest of our body, we rely on feel, experienced feel. This feel connection is exercised by our usual arms, legs and torso movements. But what of abnormal movements? This is especially challenging when ground fighting. Usually when ground fighting our vision is limited. Where are our legs versus his legs? Etc.

Proprioception is a sense that lets us perceive the location and movements of our body parts. It is mediated by specialized sense organs (proprioceptors) located within the muscles and tendons. The two classical proprioceptors are the muscle spindles and the Golgi tendon organs.

When you watch rookie fighters on the ground they are usually at a loss in this situation and miss various opportunities. It takes coaching and experience to develop the "feel" to sense-see-without-seeing" where your body parts are that you cannot see. Be aware of this problem.

This is shared with all sports and well as reaching for a cup of coffee without looking. Movements in time and space.

WHERE Question 12: Where is Urban Combatives useful? What About the suburbs? Rural areas? And now for "something completely different, as they use to say on Monty Python. A little fun, tongue in cheek. Started by an actual email I received a few years back. I changed the name to protect the innocent:

"Dear Mr. Hoochymeins, (no one can spell my last name) I am looking for suburban combatives. I see ads for urban combatives but I do not live in an urban area. I live in the suburbs. I would even settle for rural combatives as the country is closer to me than Detroit, near to my house. Can you help me?" – Ambiguous

IS THERE, ARE THERE, URBAN, SUBURBAN, AND-OR RURAL COMBATIVES?

"Dear Ambiguous, it is a bit odd yes, that there is an urban combatives in name, but there is no suburban or rural combatives. I think you should be looking for 'generic' combatives to cover all geographic problems, but the name 'Generic Combatives' is very boring and no one calls themselves that. Urban conjures up something, well, 'urban' and what? Cool? As I grew up in the New York City area and fled at age 17. Not cool or inspiring for me, but for some. Still go to any _____ combatives school near you anyway."

For me? I am a business-name-nut. Geography in a business name can mean something right away, but what exactly and for whom exactly?

Owners often use the word "global" in their title, with aspirations of eventual world reach, fame or domination? Really? GLOBAL? Or not. Others call themselves exactly what turf they want to cover. Like Piscataway Karate, they don't want to expand into Trenton, they are happy just in their little demographic, section of Piscataway. Geography involved in the title or not, business names really do count.

Professional marketeers for many styles of businesses suggest working hard on the "5 mile radius" you're in concept. "Businesses can get demographic data for a 5-mile radius around a US city from the Census Bureau." This data can include income, population, and other demographics. Businesses may use this information to:
- o   Learn local populations and trends for potential customers.
- o   Find locations close to their target market.

- Understand why some stores are underperforming,
- Identify demographic factors that contribute to their top performing stores.

Like Mr. Ambiguous, I live in the outer reaches of the ever-expanding Dallas/Ft. Worth Metroplex in north Texas. This geographic term "DFW" just continues to grow and grow, but up north here we are still surrounded by farmland and ranches. Around here, we see an occasional housing addition, then a ranch, then a strip center, then more farmland and ranches.

That breakup is what I like about the area. It's still very much country and has wide-open spaces. I am a good judge of what is rural, suburban and urban because I grew up in the thick, dense New York City area. Basically, I know city and I know country, and today's cavalier, tossed around term "urban," as in appealing to everyone everywhere, confuses me.

There was a new, small business building in the cow pasture near me. The first business in this isolated place and roadway is called Urban Nutrition. Brick wall, graffiti, art sign. You know, that ever-so-ubiquitous claw art ripping through the brick art, too.

Urban is a pretty big city name suggesting, well, what exactly for nutrition? Real, inner city … ahhh…inner city eating? Inner city, muscle growth? Inner city…vitamins? What exactly does it mean, Mister Franchise Owner? Who is it supposed to attract? Because, last I read, and for some years now, urban areas were having trouble getting available fresh food and good nutrition. Food deserts! So…copying urban nutrition plan is not much of a goal.

"Food deserts are geographic areas where residents have few to no convenient options for securing affordable and healthy foods, especially fresh fruits and vegetables. Disproportionately found in high-poverty areas, inner city, urban areas. Food deserts create extra, everyday hurdles that can make it harder for kids, families and communities to grow healthy and strong."
– Chef Google

An urban food store, with cows walking around it, in open fields, captures the very dichotomy of that name in that place.

Sure, in the next 20 years a few things will pop up all around this nutrition store I mentioned, but I will never say that it will look remotely urban, or any urban city around here. It will look suburban at best. The name sends an odd, off-mission message to us. It's just odd to have an Urban Nutrition store in the middle of a rural farmer's field on a two-lane road.

Urban, the suburban and the rural. The U.S. Bureau of the Census defines urban as a community with a population of 50,000 or more people." The dictionary says that: "Rural areas are referred to as open and spread-out country where there is a small population. Rural areas are typically found in areas where the population is rather self-sustaining. Suburban areas are references to areas where there are residences adjacent to urban areas, like those between urban and rural." - National Geographic

There is a marked difference between the three. We all know this? I see a lot of urban stuff marketed these days and, of course, even the rather ubiquitous "urban combatives" is a name dropped here and there for courses and schools. I do wonder why *that* name? I find this title curious, too. Urban Combatives. Sales pitches and advertising might be, what?

> *"Wazzup, suburb boyz? Country boyz!*
> *Fight like inner-city, urban boyz! Word!"*

> *"Fight like Boyz in the Hood."*

> *"No crime, no fights happen in the suburbs*
> *or out in the country, you stupid rednecks,*
> *just so you hicks know, down in the projects is where*
> *you really learn how to fight."*

> *"Are your punches and kicks all urbanized? Or country stupid? Run yours through our special, 'urban' filter' of urbanized special fighting that only urban thugs can do."*

Let's flip the urban chart around a bit and look at it in some opposite ways:

- o Will "Georgia Barnyard Combatives" work in Manchester or Prague?
- o Will "Harvey's Suburban Combatives" work in downtown Philadelphia?
- o Will "Jimmy Bob's Hearth of the Homeland Combatives" work in Detroit?
- o Will we ever see "Outer City Limits Combatives?"
- o Is there even a "Rural Combatives class anywhere?
- o Is there even a "Suburban Combatives class anywhere?

What? No country combatives? Seems to me urban people have no monopoly in elite fighting. Have you investigated the UFC champs for example? You know Matt Hughes is a farm boy from southern Illinois. Brock Lesnar is from Webster, South Dakota. Randy Couture is from Cornelius, Oregon. There's a long list of country folk champs. I could go on and on with this country champ list. And, champion training is conducted everywhere, not exactly an inner-city mandate.

We know what military "Urban Combat-Urban Warfare" means for today. Fighting with firearms inside cities, as opposed to say "Jungle Combat-Jungle Warfare," or "Desert Combat" or "Forest Combat." Each theater is different. At a very core it's the same, but geography varies and tactics must vary. Where you are fighting can be, might be very important.
(One thing is for sure, where you will really be fighting has no mats.)

Fighting geography is always named by professionals.

Anyway, crime and/or fights will occur anywhere. Rural, suburban, or urban. Some of the worst crimes and baddest' fights have occurred behind the barn in Idaho.

Let's talk about the martial business for a moment. Yes, fights, crime and war occur in rural, suburban, and urban areas. Indoors and outdoors. A comprehensive fighting program, appealing to the most customers, must include all these turfs in marketing. Picking one name like "urban" is actually quite limiting as far as smart business plans go, unless you are teaching in THAT very specific urban zone, teaching specific urban people, to solve their specific urban problems. Just like the military jungle fighting school teaches jungle fighters to fight in the jungle.

The marketing name of something, and advertising catch phrases, count both overtly and covertly and are major influences in the success of business. I fully empathize with the struggles to name things.

Funny thing is, many rural and suburban people that don't otherwise like the "big city," don't like their laws, politics and restrictions, some still embrace the term "urban" this or that.

Where geography worked. Geography-naming played well with **Brazilian** Jiu-Jitsu and **Israeli** Krav Maga success. In the martial arts, the "grass is always exotic and greener elsewhere than your "hometown." When in the USA, elsewhere - places like Israel, China, Japan and Brazil have the same mythological lure as Kung Fu. It seems THAT sort of green grass geography can sell well. But I don't see it with the word "urban."

Will urban work as well as in Rio or London or Mayberry RFD? I guess "urban" sounds just innocently, naively cooler to some innocent, naive people? It's not cool to me. Like I said, I grew up in New York City. Not cool at all. Good luck with all that.

WHERE Question 13: Where crimes occur.
Locations of violent crime. The following information was collected by Leonard Adam Sipes, Jr. a former police officer with a long resume of crime research while in several federal and civilian agencies.

In 2023 he examined 2022 crime in the United States. No matter what year and what country, these facts are rather universal. The top ten locations for violent crime. The article includes data from two US Department of Justice agencies, the FBI and the Bureau of Justice Statistics' National Crime Victimization Survey. Where Violent Crime Happens:

1. Your Home.
2. Public streets.
3. Parking lot and garages.
4. Elementary and Secondary Schools (not colleges).
5. Hotels-Motels.
6. Drug Store-Doctor's Office-Hospitals.
7. Restaurants.
8. Bars-Nightclubs.
9. Convenience Stores.
10. Commercial Office Buildings.

Note: Sadly, this looks like just about everywhere huh?

That spot looks safe? A study in forensic Architecture. We had a bus stop on a pretty busy street in the city where I worked. There was a horse-shoe-shaped sidewalk that cut around the bus stop and within that cut-around were some flowers and bushes. Absolutely perfectly safe-looking, busy area (busy as in lots of witnesses)...but in the daytime.

I worked major crimes for many years and was awakened in the a.m. hours by the patrol lieutenant who told me that a woman was raped behind the Eagle and Carrol Drive bus stop. My call-out job was to work the case. Before I interviewed the victim at the hospital, I made a pit stop at the bus stop scene. The patrol officer explained the scene to me and how the woman was dragged back to the cut-away and raped.

As I looked around, this perfectly safe daytime location was perfectly unsafe at night. Especially with little-to-no traffic or passerby witnesses in the middle of the night.

There was a movement years ago, now decades ago called "forensic architecture." Many architects were building edifices with counter-crime measures and not just structural safety involved. Those plans included these daytime-nighttime issues, both indoors and outdoors. For example, take Forensic Architecture in Great Britain.

> "...is a multidisciplinary research group based at Goldsmiths, University of London that uses architectural techniques and technologies to investigate cases of state violence and violations of human rights around the world."

> "Forensic architects-engineers are often called in to help determine what caused a building to collapse, a train or plane to crash, or even a car accident, particularly if some component of the machinery involved is suspected of having failed."

"...exterior building envelope leaks or vapor drive, allowing the movement of moisture through the exterior enclosure, faulty plumbing piping. Poorly installed waterproofing in interior wet areas such as bathing facilities. Failing HVAC mechanical systems. A combination of these conditions."

Now when we look up the term forensic architecture, we see the above civil suit priorities. That all sounds fantastic, but the once evolved crime prevention factor has taken a back seat.

Forensics started out as "scientific tests or techniques used in connection with the prevention and detection of crime." A serious segment branched off into pre-crime so to speak, building anti-crime, safe buildings, streets, public areas etc. with an eye for preventing crime, such as isolated bus stops, hallways, sidewalks, lighting and dangerous parking lots. I guess there are more burst water pipes and esoteric cement cracks than rapes?

There is construction today versus blatant terrorism, but what of common crime? So, take heed of where you are, where you are going and consider the light, the dark, and the "traffic" of it all. Oh, and by the way, I caught that bus stop rapist.

WHERE Question 14: Where are the Cameras? The Hawthorne Effect is a psychological phenomenon where individuals modify their behavior when they know they are being observed, often leading to temporary performance improvements. It was first observed in workplace studies at the Western Electric Company's Hawthorne Works plant, thus the name. In physics, the "Observer Effect" is the disturbance of an observed system by the act of observation. Quantum physics even has "Schrödinger's Cat," something that will give you a headache to think about, "a watched cat never boils."

So much of life now is filmed accidentally or on purpose. Like it or not. Suddenly, you find yourself in a crime or a war mess. In such a mess, are there cameras filming your mess. Will you act accordingly? Will your follow-up testimony match with what has been filmed? Or will the film prove you are a liar?

There are lots of news reports and customer-seeking lawyer webpages and ads about how terrible the police or citizens are. Body cams and street and business cameras prove that sometimes. But not always. These films can work for you too, proving your truth. In the police world, the media and news lean anti-police, anti-self-defense.

For a number of years starting in the 1980s, I was a regular guest instructor at regional Texas police academies. In the 1970s and 80s we really had no reality films, no footage like today to show cadets, just some 16mm, like L.A.P.D. or N.Y.P.D. training films. They were old-school and very staged.

If I taught or ran a police academy today, each day I would show an episode of the COPS TV show to the rookies. The show is a record of so many diverse police-citizen interactions and arrests, with cops acting at their very professional best because...they knew they would be on TELEVISION! They've been...

"Hawthorned!" Their "watched pot never boiled." So, great examples of doing - the right thing.

Since I was a kid, we were always warned that God was watching our every movement, so behave! They said the same thing about Santa Claus, but I'd worry more about God, no matter who you are. Not bad advice in our burgeoning, 1984-Big Brother world of surveillance. Nobody wants a lump of coal on Christmas. And that coal is on fire in Hell.

WHERE Question 15: Falling exactly where? Years ago, I was on a treadmill in a gym that offered a big window view of the outside parking lot. A horrendous storm was brewing. A young couple exited their car. There was a sudden, very nearby flash of lightning. A thunder boom so loud and lung shaking, that they instinctively dropped the asphalt lot! They recovered, stood and ran to the gym doors. I noticed that the man's face was bleeding from hitting the ground so hard. face not hands.

What if you suddenly have to surprise "drop?" Drop down to below cover or concealment from an explosion, or an abundance of nearby enemies, or gunfire? If you are empty-handed or your weapons are holstered, sheathed, pocketed, mounted with a shoulder sling, etc., this is a most common way to get prone. In some ways it can resemble the classic burpee exercise. To get back up you reverse the steps. Practice your "get-up" maneuvers.

"Falls are among the most common causes of injury in the United States and the most common causes of traumatic brain injuries. According to a 2016 research review, anywhere from 7 to 36 percent of falls occur on stairs. Additionally, an average of slightly more than 1 million people were treated in emergency departments annually for falls on stairs between 1990 and 2012, according to a 2017 study." - Healthline

Falling is also big topic in martial studies and arts. Either way, everyday life or training, you fall (as in fainted, tripped, slipped, struck down or tackled) on all kinds of hard floors, carpets, stairs, sidewalks, streets, grass, woods, the list goes on. Only in training do you fall on mats.)

Beginners - "Level 1" people are usually taught early on in the many martial arts have on how to fall when knocked down. You have two types of falls to worry about:

> 1: Complete surprise falls. A sudden, hardest-to-control descent. Most falls are sudden surprises. You are moving and a second later down, like an ambush, with little opportunity to respond with safety measures. Perhaps repetition training will help the unconscious response, but falling is usually very fast. "If you have a room full of soccer players (or wrestlers) and computer desk workers and go around knocking people over, you can bet the (wrestlers and) soccer players are going to be less likely to get hurt because of their superior strength, agility and coordination."
>
> - Erik Moen, physical therapist, Washington.

> 2: Predicted falls. In fights, sports and simple falling it might be possible to somewhat control a descent, after all falling is included.

Main falling response methods versus takedowns and throws taught might be described as:
- "Slapping the mat." Mat? But what about the real-world hard-ground surfaces? Who is fighting on a mat in crime and war? You slap the mat with your forearms and hands.
- "Rounding-off, rolling," curl-tuck and roll major body parts. The roll. According to paratroopers, stunt professionals, physical therapists and some martial arts instructors, there are "better" ways to hit the deck. Experts say you must save your head, elbows, wrists and knees by rolling, NOT flat mat slapping.

Alexa Marcigliano, a professional stunt woman advises, "for a safe crash landing. The moment you sense you've lost your balance, get ready to fall with bent elbows and knees. "When people panic, they become rigid. In the stunt world, we never reach out with locked arms. Bend your elbows and have some give in your arms to soften the impact." When you're rigid, you're more likely to suffer a set of injuries called "FOOSH' doctor speak for 'Fall on outstretched hand.' The result is often a broken wrist or elbow."

"When landing, aim for the meat, not bone," said Kevin Inouye, a stunt man and assistant, "Your instinct will be to reach out with hands or try to catch yourself with your knee or foot, but they are hard and not forgiving when you go down. The reach is responsible for many a bone break. Your instinct will be to stop your body as quickly as you can. But your safest route is to keep rolling — indeed, the more you give in to the fall, the safer it will be. Spread the impact across a larger part of your body; don't concentrate on impact on one area. The more you roll with the fall, the safer you will be."

"According to military research, parachute landing is the second-highest activity with a high risk of injury in the military. "The key is to not fight the fall, but just to roll with it, as paratroopers do. "The idea is to orient your body to the ground so when you hit, there's a multistep process of hitting and shifting your body weight to break up that impact,"
- Sgt. First Class Chuck Davidson, master trainer, Army's Advanced Airborne School. Ft.Bragg, N.C.

Jessica Schwartz, a physical therapist in New York City who trains athletes reports "It's almost inevitable you are going to fall, so you really should know what to do. The number one thing to remember is to protect your head. So tuck your chin and roll."

"Roll Out of It "The same concept occurs in parkour (the art of jumping and rolling around and martial arts. It is best to practice rolling out of a fall on a soft floor so that you can know what to do when the time comes." - Dr John Skedros, Orthopedic Surgery

"Landing fully on your arms in an attempt to catch yourself or 'break' your fall like a judo practitioner can break your wrists & arms." - Lawley Med Insurance

"Some martial arts, like Judo and Jujutsu, strike the ground with an outstretched arm to help absorb some of the shock. But this method is problematic off the mats, where concrete and uneven surfaces abound, and requires accurate timing to be effective."

- Micheal Grigsby, 35-year martial arts vet and owner of Fearless Falling

In many martial systems that Grigsby speaks of, mat slapping is mandatory and supposed to reduce the initial impact. I have both attended and taught in many schools in over 50 years and have seen hard, noisy mat slapping way overdone, almost like a badge of honor. Not good "street" stuff.

More roll, less slap and stop. Experts say you stop the roll with arms, you shouldn't stop the fall itself with a "slapping arm." Other systems tuck, curl and roll away. I looked for alternatives like the rolling of Silat and even Ninjitsu! And I worked on the Parachute Land Fall, a sort of vertical roll all unto its own. I never purposely slapped again.

Fall training to die? Another important point I will make is about slapping the mat. After many repetitions of falling and slapping, this creates the habit of always doing so, in training or real life. Then, when an enemy throws you down, you land, slap the pavement, possibly debilitate your limb and leave you, by your own accord laying prostrate at the feet of your enemy. If you rolled away instead? You would be escaping his next moves, one of which could be a face or neck stomp.

In training, as the trainer-uke, if you consistently fall, "slap and stop" at the very feet of your trainer, you are likely to do so in real life, leaving you laying there right in front of your enemy and his follow-ups. Be aware of this self-made trap. Get away!

When you stop your roll-away escape, or any ground maneuver escape and can't get up right away, best take a ready-ground fight position as soon as possible.

I will never slap or teach slapping for the non-mat world. Round off all your edges and roll it if possible. You will still find some old, leftover military manuals showing some traditional slapping. They have been mindlessly influenced by matted, martial arts. And some indoctrinated people will never agree and never give up these mandatory, hard, noisy slaps. Tradition! Hope they never hit the hard surfaces of the real world.

Fighting and hitting the deck accidentally or on purpose? Are you thrown or throwing? Consider these two main events:

    1: Sacrifice Falls and,
    2: Accidental Falls.

The main mission of a survival course is NOT to sacrifice fall, but to remain in a superior position, fully up or knee-high if possible. Ground Zero is a dangerous place. How you take another down is a martial art, tribal definer. By that I mean you reveal your art, sport or combatives training.

When you take down, do you want to fall with the opponent for submission tap-outs? Falling willingly with the opponent or do you try to stay up? The fall-withs or "sacrifice falls" are easier takedowns. The stay-ups require more skill with a smaller list of options.

- o Sacrifice Falls: The submission, hobby-art school of sporty takedowns use what I and some still call, and we should worry about – and survivalists try not to do - "Sacrifice falls." That is falling together, takedowns that involve the thrower using his or her own bodyweight with or without a little tripping, and "sacrificing" their own balance to take the opponent down to the mat with them.

  Willingly falling. Those seeking sport wrestling submissions need to be down on the mat. This is found in all wrestling, and is innocently, ignorantly overdone in survival, self-defense. While we emphasize remaining up, you must also have a working knowledge of these takedowns to counter them.

- o Accidental Falls. We survivalists don't want to "sacrifice fall," instead we try to remain up or at least knee high as we try to avoid the cement, glass, swamp, Astroturf, asphalt, tile, carpet, rocks, furniture and human accomplices of surrounding life. Having arrested people outdoors and indoors in urban, suburban and rural areas for many years, I-we know even untrained people, all shapes and sizes when on the floor-ground can freak-out-scrap and scramble, powered by isometrics and adrenaline and be a problem. Here, citizens and soldiers should instantly get up or resort to some or all of a ground n' pound, MMA-like, priority-mentality. (There are always situational exceptions.)

I repeat again for the record, the main mission of a survival course is not to sacrifice fall, but to remain up or knee-high if possible. Sometimes it is not possible.

Where will "they" land when punched? The "world" seemed and still seems quite cavalier about punching people, as well as shoving and tripping them down or tacking them. Perhaps from TV shows and movies. But the landing can be dangerous! The world is full of hard surfaces like flooring, the ground and furniture to name a few unknown landing zones (LZs.)

> "After being punched and falling, potential head injuries include a concussion, which can manifest as headaches, dizziness, nausea, confusion, memory problems, sensitivity to light and sound, and sometimes even temporary loss of consciousness, depending on the severity of the impact; in serious cases, skull fractures or brain bleeds may also occur, requiring immediate medical attention.'
> - John Hopkins

Many countries around the world, in both criminal and civil courts, examine a simple punch and knockdown. Where the victim landed. Did his head or ribs or arm hit a bar or table, maybe a tile floor, etc. and there were subsequent injuries? This landing may turn a simple misdemeanor assault into a felony.

"Laws covering bodily injury after being punched fall under the category of "personal injury" law, where the injured party can sue the person who punched them if they can prove the punch was a result of negligence or intentional harm, allowing them to seek compensation for medical bills, lost wages, and pain and suffering, depending on the severity of the injury and the jurisdiction's specific laws; in most cases, the burden is on the injured party to demonstrate the other person's fault." - LawInfo.com

So finally, when you hit the ground, quickly get into a position to fight! Or leave? Up to you.

Where Question 16: In a jam, where should I get to? Where should I be? The high ground. Okay, that might be a sarcastic answer, but it rings true. If downed, fight to get up as soon as possible. By higher, I symbolically refer to a better place of higher safety, and moral, ethical and legal positions.

First off, keep that phone charged up. If you are in some kind of danger. Call someone. Call the police.

What kind of jam are we talking about? That purports what solutions are needed.

- Who are you? Who is the problem?
- What is happening? What are your options?
- Where is this happening? Where can you go?
- When is this jam? Daytime. Nighttime? Season?
- How is this jam unfolding?

- Why are you in this jam?

It is easy to suggest "get to a safe place." How do you always define that? Sometimes hiding might be best? Being in, getting to, a public space with witnesses. Situational.

Being followed when driving, get to, drive to a police station is a classic suggestion, but there might not be any police hanging around the front. If so, as well it might be in the middle of the night, drive behind the station where the police cars are and often the coming and goings of officers. Fact is, I cannot define this, situation by situation, the best place to "get to." You have to "wing it."

Where Question 17: Where should I look when in a fight? The smart-ass answer is...everywhere. And well, that is partially true. You should peek everywhere, but some instructors have differing responses.

I know of a very classical system, one on the esoteric side, that suggests looking at the eyes of the opponent, so to see his... soul. But no. You do have to glance at their face and eyes. Here's when and why. In these you will be looking, scanning, or as the eye pros call it, "gazing" at a number of places at a number of times and "distance locations." "Optometrist Dr. Google" reports:

> "In eye movement studies, "gaze" refers to the point where a person is visually focusing their attention, essentially the location on which their eyes are fixated, and is often measured by

analyzing the duration and pattern of eye fixations on a specific area during a given task, providing insights into cognitive processes and where someone is directing their visual attention."

Especially in the Collision 1, stand-off, pre-fight range, you are sort of looking and gazing everywhere, the location and at the suspect-person (or persons). And you can glimpse the eyes. In fact, in police work there are pre-fight, interview moments when the suspect looks-checks you out and or over-checks the gun you are wearing. A bit too much. Is he planning to take your gun? Or does he just like guns?

We are taught to say something like, *"Sir. Sir! Can you stop staring at my gun and look at me when we are talking?"* This alerts the suspect that you know what he is doing and thinking. Otherwise, his darting eyes searching the area, tips us off that he is:

a) looking for an escape path or,

b) worried about witnesses seeing him jump you.

So, there is a time and place to check into his eyes, usually in that pre-contact, Collision 1 showdown arena. But when the action starts do we just look at their eyes? No, not so much. Not never, just not too much.

In the pre-fight geography, you will hear folks in the police, military and civilian gun world suggest, "watch the hands, it's the hands that will kill you." You can't stare at the hands! If possible, worry about his hands through the whole fight. In terms of weapons, His hands will reach for their:

1) primary carry site (quick draws on their body like the beltline and pockets),
2) secondary carry sites (digging out more hidden, back-up weapons), or
3) tertiary carry sites (lunging to reach for off-the-body-weapon sites.

Yes, worry about their hands, but you cannot conduct yourself in or out of a confrontation over-staring down at the other guy's hands. Hand movement can be distracting and hypnotic.

The fight starts. The suggestion is to look at the center of the chest just below the clavicle. This picture maximizes your peripheral vision; you can see the shoulders and arms move around on down to about the tops of his thighs. If you train, these incoming movements are predictors of strikes, tackles and kicks.

Another advantage to this high-center-chest-vantage point is you take the "face out of the fight," so to speak. The face (and the voice) of the enemy can be a serious distraction to the survival tasks at hand.

Face wise, he may look worse than he can fight (scary?) and fight worse than he can look (not scary?). Collision 2 and onward, it's now time to create and handle events and not be falsely distracted. In other words, he may look like Jack Reacher (the Alan Richson version, not Tom Cruise) and fight like Peewee Herman?

Still, you need to gaze around a bit. Once in closer quarters, like Collisions 5 and Stop 6, you might see little but feel more, your "seeing' becomes feeling and we have previously studied the subject of

Proprioception - physical seeing where you and your enemy's body parts are.

Seeing well and sharply is part of performance success in sports, policing and the military. Two books I strongly suggest you read...

The WHERE Summary. So where can you forecast trouble? Where are you in your education, your training, your life?

The Department of Homeland Security, primarily interested in terrorism, suggests you look for suspicious activity at these places, "government buildings, religious facilities, sports/entertainment venues, high-rise buildings, mass-gathering locations (e.g., parades, fairs, etc.). Schools. Hotels. Homes. Theaters. Shopping malls. Bridges. Public transportation." Well! I think that is just about everywhere, isn't it? Continue to ask and investigate the many small and big where questions...

The Where Question Review.

Who are you within this where category?
What are you within this where category?
Where are you within this where category?
When are you within this where category?
How are you within this where category?
Why are you within this where category?

Chapter 6:
The When Question and Confrontations

"You always pack your parachute before you jump out of the plane." - old military adage

When do you feel the safest from crime and war? Conversely, when are you most vulnerable to them? In your daily life, how do the places you visit fluctuate in safety depending on the time of day, week, month, or year?

When might someone lie in wait to ambush you? At what moment? Day or night? During lunch breaks? At the dinner table? On early morning walks or runs? While stopping for gas? When does the unexpected strike? A stealth raid on force protection? An invasion? The dreaded ambush!

This chapter centers on time. Time is just a measurement. Crime rates and details that include times change year-to-year around the world. Pew Research concludes with many sources that:

> "The two primary sources of government crime statistics – the Federal Bureau of Investigation (FBI) and the Bureau of Justice Statistics (BJS) – paint an incomplete picture."

You see, no one, not even Big Brother and Uncle Sam can precisely predict when you will be confronted, ambushed and attacked. You don't decide this, we don't decide this, your attacker does. It's their ambush. Criminals do. The enemy does. We are on their timetable. We can only plan and "make ready." It will all be some level of their verbal-interview contact and/or physical ambush.

Enemy forces invade when they want to. Look at Nazi Blitzkrieg of WW II. Pearl Harbor, Hawaii in 1941. Look at the October 7, 2023, surprise Hamas raid into Israel.

In crime, in the USA, we are used to hearing expressions like "you have a "1 in 10,000 chance" of being assaulted, etc. That sort of statistical scrambling might be comforting, but the old line "hope is not a strategy" is true. And when you are actually *thee* one being attacked, then you become "1 in 1." Not 1 in 10,000.

We can only say here for sure when such an attack will happen. Like…NOW! It will be now to you when it happens. The surprise moment it happens. Use your planning-forecasting skills to determine who, what, where, when, how and why, as in WHEN locations can become dangerous. I hope these following questions will help you think and plan.

WHEN Question 1: The Before, The During, and The After. One of the most important yet often overlooked concepts in preparing for life, survival, peacekeeping, crime, and war is understanding that every event has three stages: before, during, and after.

This applies to everything in life when you break it down. Consider viewing your decisions through this lens, like looking through a spyglass. Whether it's buying a car, purchasing a house, getting married, practicing self-defense, voting, punching someone in the face, or even shooting a bad guy, every action follows this pattern:

- o Before – The planning phase, including motivations and preparation.
- o During – The execution phase, involving mental and physical actions (when, by the way, most martial training tends to focus, oblivious to the before and after)
- o After – The outcome phase, where you face the good or bad consequences of your actions. Understanding this framework can help you make better choices and anticipate the full impact of your decisions. Remember the goals,
  - stay out of the hospital.
  - stay out of jail.
  - stay out of civil court.

WHEN Question 2: When must I be alert? Life is messy. People worry about verbal confrontations with coworkers or bosses, challenges in the workplace and home on up to physical confrontations in crime and war. Instructors demand people remain, "Always be on the alert!" expecting problems.

So easy to proclaim. But living a paranoid or even slightly paranoid life can scorch your insides from bodily stress fluids. The famous Mayo Clinic discussed such consistent, cortisol-adrenaline overdoses:

> "When the natural stress response goes wild. The body's stress response system is usually self-limiting. Once a perceived threat has passed, hormones return to typical levels. As adrenaline and cortisol levels drop, your heart rate and blood pressure return to typical levels. Other systems go back to their regular activities.

The long-term activation of the stress response system and too much exposure to cortisol and other stress hormones can disrupt almost all the body's processes. This puts you at higher risk of many health problems, including:

- Anxiety.
- Depression.
- Digestive problems.
- Headaches.
- Muscle tension and pain.
- Heart disease, heart attacks, stroke.
- Sleep problems.
- Weight gain.
- Problems with memory and focus.
- That's why it's so important to learn healthy ways to cope with your life stressors.

Can you walk around "on high alert" 18 hours a day as some instructors suggest? Prepared for daily problems, arguments or attacks? Prepared for whatever it is in life that worries you? How can you remain stable? How can you predict when to worry and when not to? How can you prepare for the dangerous ambushes in life and remain internally unscorched?

Since interviews-arguments and ambushes occur every day, and I like to call the dialogue part of those meetings - the "scripts of life," like it or not we typically, predictably follow a rehearsed, familiar "talking trail" in these scripts. Jump off the path! I hope you can extrapolate this Ws and H subject matter into the interview and ambush pictures of life.

- Who am I to be so alert all of the time? Even some of the time?
- What should I be alert for? What level?
- Where should I be alert?
- When should I be alert?
- How do I act when alert?
- Why should I be alert at certain times?

WHEN Question 3: When are the times when I must likely be attacked, involved in some sort of crime? Times of the day. It's the "when there, when not there" line. The old running phrase was, "Nothing good happens after midnight." In seminars I repeat this line, but add, "I am in my 70s. Nothing good happens after 7:30."

Everyone laughs. But overall, when is it safe to be "out and about?" A consensus of crime and time experts agree...ahhh, well, they don't agree actually. I have spent a considerable amount of time trying to crunch numerous sources and the numbers vary city by city, state by state in the USA and the world. So much so I hate to pontificate on any solid numbers here. Getting something of a grip on this when question:

- Criminologists all agree that summer months in countries produce more crime. Warmer. More people out and about.
- Some believe that December is also a busy crime month, as people need money for the holidays, and victims come out of the woodwork shopping.
- Overall, crime in metropolitan areas (and I will assume in suburban and rural areas) was generally lowest in the late night and early morning hours, especially between 4-7 a.m. Criminals, terrorists and soldiers do and will sleep at some point. When we did surveillance on thugs, we hung on until their lights were out, so to speak, and we decided-figured that we "put them to bed."
- The National Library of Medicine has deduced that most gunfights occur in the hours of darkness.
- Robberies increase around "payday" operations. Specific things happen at specific times and places. I recall my times in the military police, both patrol and investigations, that a lot of bad things happened on paydays. Every two weeks like clockwork, hookers, muggers, robbers and drug dealers traveled to the cities beside and near our bases. We treated paydays as special missions.

- Some cities hate tourist seasons for these same and compounding reasons. The State of Florida, USA is inundated with college spring-breakers. Crime increases.

I will conclude this cloudy collection of advice by simply saying, look for safety and crime specifics in your areas at home, work and travel.

WHEN Question 4: When are you the weakest at your decision making? When are you most likely to make poor, impulsive, or ill-considered decisions? The answer: when you are tired, distracted, afraid, angry, or in pain. These moments of mental and physical weakness can significantly impact your judgment. Many factors can impair decision-making, but these are some of the most critical. Over the years, as part of my training courses, I have developed three key areas of focus:

        1: Anger Management
        2: Fear Management
        3: Pain Management

I am not a psychiatrist. Likewise, most martial artists and combatives instructors are not trained psychologists. And while I have attended a criminal profiling school, I do not claim to be an FBI profiler.

My role is not to diagnose but to guide. Rather than pretending to be experts in these deep psychological fields, we should direct people to qualified professionals when necessary. The following three essays explore these critical moments of vulnerability and how they affect confrontations and decision-making.

Essay 1: Anger management in violence, self-defense, and combat. Anger management is one of the three key "managements" my course identifies since the 1990s as priorities for awareness, training, and performance in dealing with violence, self-defense, crime, and military combat. Anger is present:
   1. Before a confrontation or combat.
   2. During the struggle.
   3. Even after the conflict has ended.

According to standard psychology:
"Anger management is the process of learning to recognize signs that you're becoming angry and taking action to calm down and deal with the situation in a positive way. It doesn't aim to suppress or ignore anger but rather teaches how to express it appropriately. Anger is a natural, healthy emotion when managed correctly. While some may learn anger management skills independently through books or resources, many find structured classes or professional counseling to be the most effective approach."

Anger is a powerful emotion that plays a significant role in conflict and survival. It is a leading cause of fights and a major factor in domestic violence. It is deeply intertwined with adrenaline and its physiological effects.

Psychological stressors naturally come into play when dealing with anger, making this a field where professional counselors can provide valuable insight. If you are not the instigator of a confrontation or act of violence, then you are being confronted or attacked.

A common emotional response to such situations is anger. However, losing complete control, transforming from Bruce Banner into the Hulk, can be detrimental to both your physical survival and your legal standing. The goal is to control anger, not be consumed by it. Use it. Don't lose it. Don't abuse it.

Managing Your Anger in Confrontations. If you are not caught completely off guard and have a moment to react, consider the following steps. These are just a few techniques that can be effective when time allows. Recognize the signs of rising anger.

- Clenching your fists.
- Feeling flushed or sweating.
- Rapid heart rate.
- Trouble breathing.
- Feeling like you might explode.

Use psychological and physical techniques.

- Humor can sometimes help defuse tension. Take note of the political comedians. They beat the politics drum, then break it up with comedic remark, that keeps them from full-bore complaining.
- Some people smile under pressure, which may trigger calming biochemical actions.
- Employ physical relaxation techniques.
- Learn to "center" yourself when possible. Take slow, deep breaths, focusing on your breathing. (much more on this later.)

- Tense and release muscle groups, start with your hands, legs, back, and toes.
- Repeat a calming phrase to maintain control and confidence.
- If necessary, remove yourself from the situation.

Harnessing Anger for Survival. Zero to sixty stuff. As with the "zero to sixty" material previously discussed, some "mileage" in anger is productive. Again, I quote my old friend David Hackworth who said, (and paraphrase),

> "I want a soldier in the trenches with me, who when doing the dishes at home, when he accidentally hits his head on a cabinet door, cusses, growls and gets angry. Maybe greatly desires to punch the cabinet door. But he controls it. Not crazy angry. This guy will get angry enough to kill the enemy. Not crazy, stupid angry, just angry…enough. I don't want some peaceful, yoga hippy in the trench with me."

Controlled anger, rather than reckless anger, can be a source of strength and resilience. Anger can be slow-burning or explosive, and your lifestyle can influence how it manifests. In my personal experience, when engaging with violent suspects, I have found that being half-angered and half-adrenalized is the optimal state. Achieving this balance requires self-control and a lifestyle that prevents destructive, simmering rage.

In the realm of survival and self-defense, it's important to keep a "wolf" inside you, controlled, but ready. Feed it. Feel it. Occasionally rattle its cage in your training. One day, you may need to set it half-loose to survive. 30 mph, not crazy 60. Self-reflection Qs on anger...

- Who are you when you're angry?
- What triggers your anger?
- When does anger surface in your life?
- Where do you experience anger the most?
- How does anger affect your decisions?
- Why does anger manifest in you?

Understanding these Ws and H questions can help you harness anger as a tool rather than letting it control you and your decision-making.

Managing Your Opponent's Anger. Back to the "you-them duality." Anger management isn't just about controlling your own emotions; it also involves managing the anger of your opponent. In pre-fight situations, your actions, looks, words, or even silence can influence your adversary's emotional state. You may choose to de-escalate or escalate, and in some cases, use a scare tactic to shift the dynamic.

Effectively, this is about managing their anger, not just your own. Entire books and academic courses explore this subject in depth. A short seminar, a martial arts class, or even a guide like this can only scratch the surface. What you say, how you say it, your facial expressions, and your posture are possible control measures. For deeper understanding, refer to experts in psychology, conflict resolution.

Essay 2: Fear management in violence, self-defense, and combat. Fear management is one of three critical priorities - alongside awareness and training - that my course emphasizes for dealing with violence, self-defense, crime, and military combat. History has shown that even the most formidable armies and skilled fighters have been defeated by the shock of sudden ambushes. Understanding and managing fear is essential for performance in high-stakes situations.

I offer here a few expert strategies to dissect, compartmentalize, and manage fear. While these methodologies apply to a range of fears and phobias, our focus will be on fear in the context of violence,

fighting, crime, and combat. The fear we examine can occur:
1. Before a fight or struggle
2. During a fight or struggle
3. After a fight or struggle

The study of fear spans a broad spectrum, varying by an individual's lifestyle, job, or mission. Fear is not a predictable, gradual progression; it fluctuates unpredictably. We will not cover long-term psychological, post confrontation, conditions stemming from fear.

The military and law enforcement perspective on fear. Contrary to popular belief, not all soldiers or officers freeze in fear or lose control in combat. Some become overwhelmed, yes, but others experience heightened stimulation and excitement. Modern military and law enforcement recognize these differences and strive to cultivate units of highly capable, stimulated personnel. Managing fear and adrenaline is key to performance in high-stress environments. Training and selection processes help identify individuals best suited for the front lines.

The military creates-manages-projects fear into the enemy through a combination of physical force, psychological operations, and intimidation tactics designed to demoralize and disrupt their ability to fight. This goal, also known as deterrence, involves producing in the enemy's mind the fear to sustain or to attack.

For examples, advancing Scottish bagpipes intimidated foreign troops through history. In 1982 Falklands War (or Malvinas War) was a short, undeclared conflict between Argentina and the United Kingdom over the sovereignty of the Falkland Islands and its dependencies. The threat of deploying the Gurkhas', with their fearsome reputation played a significant role in frightening the Argentinian troops leading to panic and rapid surrenders. The ROK Marines of South Korea instill fear. Military history is replete with such fear management campaigns.

The two innate fears. Everyone is born with two inherent fears: falling and loud sounds. The rest are learned through experience.
- Fear of Falling (FOF): A natural response found in most humans and mammals, distinct from acrophobia (fear of heights).
- Response to Loud Noises: The "acoustic startle reflex" triggers a fight-or-flight response, signaling potential danger

.

Fear-related responses are shaped by experiences, training, and environmental cues. For instance, firefighters learn to gauge the intensity of a fire by the sound of its roar, snaps, and pops.

Fear as Pattern Recognition. Fear responses, often mistaken for "gut instinct," are actually complex neurological calculations. The brain rapidly assesses patterns based on prior experiences, a process known as patternicity, the ability to recognize patterns, a hallmark of intelligence.

> "Our brains are belief engines: evolved pattern recognition machines that connect the dots and

create meaning out of the patterns we
perceive." - Dr. Michael Shermer

Understanding fear: "No Fear" vs. "Know Fear."
Martial arts magazines, courses, schools and instructors once promoted the idea of "No Fear." However, research forced a shift toward "Know Fear," the acknowledging that fear is natural and can be harnessed for preparation and survival. Books promising to "conquer fear" mislead readers as fear cannot be entirely eliminated, but it can be managed and used as a motivator.

Even professionals experience fear. Frank Sinatra once admitted that he trembled for the first few seconds before stepping on stage but quickly regained control. Even the most dangerous special force operators readily state they have combat-mission fears. Fear, when understood, can be channeled into performance.

The Evolution of the Fight-or-Flight Response. The classic fight-or-flight model, coined by Dr. Walter Cannon in 1929, has long dominated discussions on human responses to threats. However, modern research has expanded this view. The Four Fs: Fight, Flight, Freeze, Fright:
- 1. Fight – Actively resisting or countering the threat.
- 2. Flight – Escaping or retreating from danger.
- 3. Freeze – Momentary immobility, assessing the situation.
- 4. Fright – Paralyzing fear, often leading to total inaction.

Fear of the "angry bear" analogy used by psychologists, where an individual either fights or flees upon encountering a bear, ignores the common response of freezing in place.

Freezing may result from fear, but it can also be a reaction to sensory overload, unrelated to fear itself. Thus the expression, Frozen from disbelief, in what is happening before them. Not cowardice, just shock. Modern experts propose updating the response model:
> "We propose the adoption of the expanded and reordered phrase 'Freeze, Flight, Fight, or Fright' as a more complete and nuanced alternative to 'Fight or Flight.'"
>
> - Psychosomatics: Journal of Consultation and Liaison Psychiatry

This revised framework is now widely used in medical and psychological fields, including speech therapy for stuttering and PTSD treatment for combat veterans.

Tonic immobility: The extreme freeze response. In extreme danger, some individuals experience tonic immobility, an involuntary, coma-like paralysis. This survival mechanism, seen in animals playing dead, may serve as a last-resort defense against predators.
> "Tonic immobility is a reflexive and involuntary reaction, elicited under the perception of overwhelming danger, characterized by profound motor inhibition and relative unresponsiveness."

— Neuroscience & Biobehavioral Reviews, 2017

A fear essay: The small hairs on fear. There are also many small hairs to be split on this subject. As an aside, about small hairs and their standing on end. Dr. Wikipedia reports, "When the Sympathetic nervous system is triggered you will experience an increase in heart rate and blood sugar levels, sweatiness, dilation of pupils and that feeling we all know, hair 'standing on end'. In short, your body becomes 'stimulated,' in order to defend the body against attack. The arrector pili muscles, also known as hair erector muscles, are small muscles attached to hair follicles in mammals. Contraction of these muscles causes the hairs to stand on end, known colloquially as goose bumps (piloerection)."

Another fear essay: The Incredulity Response: Frozen from disbelief. This is important to know. Naive TV news viewers complain, and subsequently citizens complain when they see films of citizens ignoring a vicious assault occurring on the street, or say, in a pizza customer line before them, saying they seem to freeze or purposely ignore the crime! How uncaring! Inhumane!

The "incredulity response? Dr. John Leach, author of Survival Psychology, teaches an advanced course in survival psychology at Lancaster University in England. Leach has another name for some freezing (and for people who seem to ignore or appear cowardly when crimes happen before them).

It's called the incredulity response because people simply don't believe what they're seeing. So they go about their business, engaging in what's known as "normalcy bias." Under-reactors act as if everything is OK and underestimate the seriousness of danger to themselves or others.

Some experts also call this "analysis paralysis." People lose their ability to make decisions. Leach says that "of the vast majority of us ... (80 %) in a crisis, most will quite simply be stunned and bewildered. We'll find that our reasoning is significantly impaired, and that thinking is difficult. It's OK, leach says, and "it doesn't last forever. This has nothing to do with courage or cowardice, just a shocking lack to read, grasp and react to what is going on."

Yet another fear essay: Old school hypervigilance? It's difficult to separate hypervigilance from fear:

> "Overestimation of a threat, Hypervigilant individuals are constantly on the lookout for dangers that are either unlikely or exaggerated. This might include isolating themselves to avoid potential attacks, sitting near an exit to quickly escape, or positioning themselves with their back to the wall to prevent anyone from sneaking up behind them.
>
> Hypervigilance occurs when our natural fight-or-flight instinct goes into overdrive. Those who experience hypervigilance live in a constant state of anxiety. While it isn't a diagnosable mental health condition, it is a

> common feature of a wide range of physical, psychological, and cognitive disorders."
> – Very Well Mind

"Excessive hypervigilance, especially when triggered by a broad array of danger cues, can become maladaptive. It has been linked to negative cognitive and behavioral effects such as attentional bias, memory impairment, and difficulties with emotional regulation. Furthermore, hypervigilance plays a role in maintaining anxiety disorders and post-traumatic stress disorder (PTSD). Anxiety heightens hypervigilance, which leads to more threat detection, thus perpetuating anxiety and hypervigilance. On a physiological level, excessive hypervigilance can cause autonomic arousal and overactivity in brain regions associated with threat detection." – National Library of Medicine

One approach to managing hypervigilance involves focusing on the 5 Ws and H questions. This helps individuals compartmentalize and understand the difference between "probabilities versus possibilities." Hypervigilant individuals must learn to downplay possibilities, as an endless list of "what ifs" can fuel paranoia and exhaustion.

Before the end of the 20th century, hypervigilance was often associated with the "incredulity response," where individuals would freeze in the face of overwhelming circumstances, not just fear. With the rise of PTSD in recent decades, further distinctions of hypervigilance emerged. An early, notable one was "Hypervigilant-Freeze."

This concept was introduced to me during an FBI Defensive Tactics course years ago, with instructors using the example of a speeding train. Imagine you're suddenly on train tracks, and a speeding train is rapidly approaching. Your brain knows you and decides, "I know you and you can't outrun this!" So, you freeze. You surrender to that negative message. But the instructors explained that if you train your body, with sprints or running exercises, for example your brain will respond differently, "I know you. You're fast. You've got this! Go!"

This idea resonated with me then and still does. However, many modern definitions of hypervigilance, particularly in the context of PTSD, have overshadowed older concepts, and it's extremely difficult if not impossible to search-find recent research on examples like the "train track" scenario.

The challenge of training individuals to manage fear is significant. Some people are genetically predisposed to struggle in such situations, presenting a real obstacle. These individuals must be identified and, in some cases, reassigned away from front-line roles in police, military, fire, or EMT services. But what about the average citizen who suddenly finds themselves facing a fearful challenge?

To overcome fear, both individuals and groups must first confront their specific fears and recognize how destructive they can be to survival and success. Answers to the 5 Ws and H questions can act as a roadmap. Mapping out the terrain helps forecast and clarify what lies ahead. Fear is essential for survival, prompting us to avoid danger or prepare for problems.

It's vital to have a healthy amount of fear, it should never be fully eliminated.
- Who are you, when are you afraid? When do you think are you hero or a coward?
- What makes you afraid?
- Where are you afraid the most? The least?
- When are you afraid? Where are you afraid?
- How does fear affect you?
- Why are you afraid?

Conclusion: Managing Fear for Success.
Understanding fear is crucial for those in combat, law enforcement, and self-defense. Instead of denying fear, we must recognize, analyze, and manage it to enhance performance. Whether through training, experience, or innate resilience, the ability to handle fear determines success in high-pressure situations.

There's also a process called fear extinction, defined in the scientific literature as a decline in fear responses due to multiple "non-reinforced" exposures.

"Fear extinction is not forgetting a memory; it's actually forming a new memory," said David Connor, PhD, a postdoctoral researcher in the lab of John A. Dani, PhD, and chair of the department of Neuroscience, University of Pennsylvania Health System 'It's learning a new safe association between the context or whatever cues were initially associated with danger. During extinction, you have a second form of learning happening that is competing with the original fear memory. For example, if someone is in an accident while driving through an intersection, they may feel afraid of getting hit when going through that same intersection later. It may be that the feelings of fear can only be put behind them by driving through that same intersection several or many times."

Dr. Steve Graff with Penn Medicine News advises, "Know fear, and you'll control it. Ignore fear, and it will control you."

Essay 3: Pain management in violence, self-defense, and combat. Pain management is one of the three key "managements" that my course identifies as crucial for addressing violence, self-defense, crime, and military combat.

Pain. Yours and theirs (always remember the duality, this is about both your pain and the pain you inflict upon another. You also manage their pain.

There are plenty of pain management courses and books, but almost all of them focus on EMT services, emergency room treatments, and long-term care of pain caused by accidents, aging, or sports injuries. Many, if not most, internet searches on the topic lead to these areas. However, there is little information available on the sudden pain incurred in violent encounters, such as fistfights, stick fights, knife fights, or gunfights. Adrenaline helps pain management, and even tactical breathing can be beneficial. We'll discuss these breathing methods later in this book.

Most experienced, serious contact sports athletes, military personnel, and law enforcement professionals understand the importance of preparing themselves mentally and physically for impending pain. These practices serve to inoculate you for the inevitable conflict.

"Your brain's wiring could be an indicator of your threshold for pain. One study reveals a correlation between a person's sensitivity to pain and the thickness of the cortex in the brain. Other studies suggest that less gray matter may also be linked to higher pain sensitivity." – The Natural History of Pain

What is pain tolerance? Dr. Google defines it as "the level of pain that a person can bear." For about 50 years, I've collected personal reports and memories from individuals who've been hit, stabbed, slashed, or shot. From my years in police work, training, and research, I've found that many factors influence how people react to being bent, twisted, punched, kicked, slashed, stabbed, or shot.

Many people who have been stabbed or slashed by edged weapons often don't realize it at first. If they did feel it, , they often describe the sensation as a punch or hammer-fisted blow if they didn't see the knife. Similarly, people who are shot frequently describe the feeling as being hit by a fast-pitched baseball or a baseball bat. Some even describe it as a slight burn, akin to a cigarette burn. Others report the sensation as a simple "tap."

But taking a punch to the face or body? As you can see in MMA, boxing, or even on YouTube, sometimes there's a dramatic reaction ("ass-over-tea-kettle"), and other times, there isn't such a fall. It depends largely on the positioning within the incident.

Adrenaline mixed with anger can sometimes help you push through. I don't need to remind you that training, along with occasional "training pain" experiences, is invaluable. Mentally and physically steeling yourself makes a difference. For example, exhaling before torso impact helps tighten the torso muscles, "steeling yourself" one of many physical methods I teach in my courses.

Some martial artists and sports experts offer valuable advice on this subject. It can also guide you on which body parts to target and protect. Study common injuries like cramps, concussions, muscle tears, bone fractures, knife wounds, and gunshot injuries. Several factors influence the complex communication system between your brain and body. Here is a compilation I collected:

- Genetics: Research suggests that your genes can affect how you perceive pain. They may also influence how you respond to pain meds.
- Age: Older individuals may have a higher pain threshold, though more research is needed to understand why.
- Sex: For reasons not entirely understood, females tend to report more intense and longer-lasting pain compared to males.

- Chronic illness: Conditions like migraines or fibromyalgia can alter your pain tolerance over time.
- Mental illness: Individuals with depression or panic disorder often report heightened pain sensitivity.
- Stress: High levels of stress can make pain feel more intense.
- Social isolation: Lack of social support may increase the experience of pain and decrease pain tolerance.
- Past experience: Pain inoculation. Previous experiences of pain can shape how you respond to future pain. For example, people accustomed to extreme temperatures may have a higher pain threshold. Conversely, someone who's had a bad experience at the dentist may react more intensely to minor procedures in the future.

"Pain inoculation" is a psychological term referring to a technique used in Stress Inoculation Training (SIT), where individuals are taught and practice coping skills to manage and build resilience against future pain or stressful situations. The term uses the analogy of a medical vaccine, which introduces a weak version of a virus to build resistance, to explain how exposure to manageable stressors (e.g., mild pain in an experimental setting) combined with coping strategies can "inoculate" a person against more intense or chronic pain in the future. The process of pain inoculation typically involves three phases:

- *Conceptualization:* Education about the nature of pain and stress, emphasizing its psychological dimensions and identifying personal stressors and current coping mechanisms.
- *Skills Acquisition:* Training in various coping skills, such as relaxation techniques (e.g., progressive muscle relaxation), deep breathing, problem-solving strategies, and cognitive restructuring to manage upsetting thoughts.
- *Rehearsal and Application:* Practicing these coping skills during exposure to controlled, minor noxious stimuli (like the "cold pressor" task, where a hand is placed in cold water) or imagined stressful scenarios, and eventually applying them to real-life situations.

The goal is to change a person's perception and reaction to pain, giving them a greater sense of control over their pain experience. This approach has been explored for managing both acute (e.g., post-surgical pain) and chronic pain, with varying degrees of effectiveness depending on the context and intensity of the pain. Your upbringing and learned coping strategies can influence how you perceive and react to pain.

There are, of course, pain tolerance tests, often using a scale of 1 to 10 to assess how much pain a person can endure. But after reviewing some of these tests, I find they remind me of the horrifying medical experiments conducted in Nazi concentration camps!

"You can play with hurt, but you can't play with injury." – Professional NFL Champ

- Who are you when in pain?
- What causes your pain?
- When do you experience pain?
- Where do you feel pain?
- How does pain affect you?
- Why are you in pain?

The three managements and brain management summary. The brain does more readily, what it does more often. Neurons that fire together will wire together. Practice makes perfect (whatever that practice is). This process is called neuroplasticity. It's how our brains learn, adapt, and sometimes develop unhealthy patterns. For one example, if you often feel jealousy in your relationships, your brain might start to see your partner's interactions with others as a threat, even without real cause. Over time, this can lead to obsessive thoughts, anxiety, and controlling behaviors. Get a grip. Make fast and slow decisions while grappling with anger, fear and pain.

WHEN Question 5: When are the dangerous times of day and the days you drive. Crashes and Road Rage? All confrontations of sorts. Some good international advice:

"The terms road rage and aggressive driving are often used interchangeably, but there is an important distinction between the two, according to the National Highway Traffic Safety Administration. Aggressive driving is often used as a label for unsafe driving behaviors, like speeding or tailgating, that could compromise people's safety and lead to a traffic violation. Road rage incidents include driving behavior that can escalate to yelling, angry gestures or violent acts.

Outside of raised voices and rude gestures, road rage is a criminal offense. Aggressive driving and road rage warning signs.

No two drivers are the same, so the way they express their frustration might be different. Here are some signs that a driver might be upset, according to the New York Department of Motor Vehicles.

- Tailgating.
- Speeding.
- Cutting other drivers off.
- Running red lights or stop signs.
- Swerving in & out of heavy traffic.
- Not using signals.
- Preventing other cars from passing or changing lanes.
- Flashing bright headlights.
- Passing a stopped school bus.

When aggressive driving escalates into road rage, you might see a driver:
- Throw objects at a vehicle.
- Scream at other people or make angry gestures.
- Ram another vehicle.
- Sideswipe another vehicle.
- Force another vehicle off the road.

But it's important to stay in control of your emotions behind the wheel. Here are some tips to help you avoid succumbing to aggressive behaviors. Again, a compilation:
- Give yourself time to get where you're going. Before you ever get behind the wheel, create realistic expectations about your travel. If you need to be somewhere at a specific time, make sure you factor in expected traffic or possible delays.
- Don't drive while experiencing intense emotions. If you had a frustrating day at work or got in a fight with a dear friend, you might

not be in the best headspace to get behind the wheel. - DefensiveDriving.org.

- "If you're already angry or upset, don't drive. Put off non-essential trips until you calm down. If you have something scheduled that you can't miss, ask a friend or family member to drive, or consider calling a rideshare service or taking public transit.
- Build positive driving habits. Don't speed through traffic or weave between vehicles. Avoid cutting other drivers off and making rude gestures or remarks. The Federal Highway Administration says these behaviors are some of the most dangerous.
- Only use your horn if necessary. Car horns are just one small hand movement away, but they're primarily designed for emergency use. Lay off the horn and show them a little grace.
  - Commercial Vehicle Safety Alliance
- "Be understanding of other drivers. Before you allow your frustration to build, try putting yourself in the other driver's seat. Chances are, you've accidentally sat at a green light for a few seconds or forgotten to signal ahead of a lane change. If you remind yourself that other drivers aren't perfect or out to get you, it can help prevent you from responding aggressively.
  – Aceable
- You can only control your own behavior and driving decisions, but you can drive defensively and take specific actions to keep yourself safe from other angry drivers. Here are some more

best tips for when someone around you is acting recklessly.

- Remove yourself from the situation. The California DMV stresses it's important to avoid drivers exhibiting dangerous behavior however you can. Dangerous behavior includes tailgating, aggressive braking or swerving. Create distance between yourself and the aggressive driver by getting over to let them pass or slowing down and watching them speed off into the distance.
- Ignore them if they're being rude. Ignore angry gestures, advises the Texas Department of Insurance. You don't want to escalate the situation, and you never know what could set someone off and make them become violent.
- Call for help if necessary. If you're an angry driver's target, call 911 immediately and drive to a public location or your nearest police station, says the Seattle Times. Even if an aggressive driver is only targeting one vehicle, they're a threat to everyone on the road. Calling the police helps ensure everyone's safety.
- You might "long-distance apologize" after a mistake. A waving of hands, even "lipping" the word "sorry." No driver is perfect, so always apologize when you make a mistake. If you inadvertently proceed without the right of way or get too close to the car in front of you, a friendly wave and smile can help keep everyone calm. - Andy Pilgram. Traffic Safety Education Foundation.

- When driving? Keep your windows mostly closed or fully closed, and your car doors locked.
- Here's an idea. Don't pull over and stop when the other guy is acting like an idiot. We realize though that some of you are stuck-stopped at a red light with the idiot nearby. You are left to the steps, situations and rules of self-defense.

What happens to us?

Before we even cognitively process the situation, our brain goes into "fight or flight" mode and releases hormones, such as epinephrine and adrenaline, which cause a cascade of other reactions:

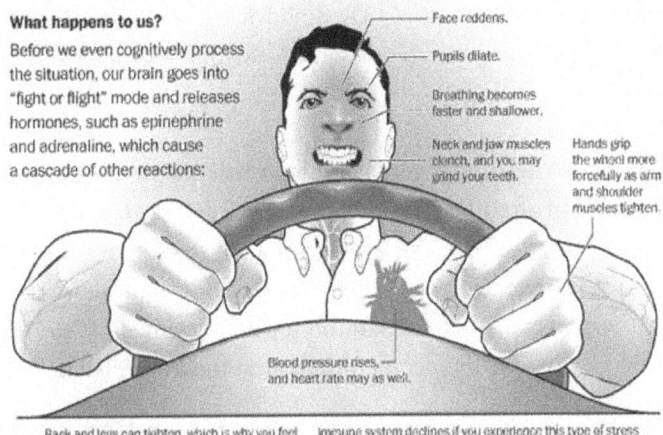

- Face reddens.
- Pupils dilate.
- Breathing becomes faster and shallower.
- Neck and jaw muscles clench, and you may grind your teeth.
- Hands grip the wheel more forcefully as arm and shoulder muscles tighten.
- Blood pressure rises, and heart rate may as well.

Back and legs can tighten, which is why you feel stiff getting out of the car after fighting traffic. Immune system declines if you experience this type of stress daily, in addition to potential cardiac and circulatory problems.

What triggers it?

Here are behaviors that were mentioned most often in Wickens's 2013 study, which tracked more than 5,500 gripes over nine years posted to Roadragers.com:

Number of complaints by category

Category	Count
Weaving/cutting	2888
Speeding	1563
Hostile display, flipping the bird	1325
Tailgating	1282
Lane usage	734
No turn signal	713
Erratic braking	626
Running red light, not stopping for pedestrians	469
Blocking	466
Violent display	421
Slow driving	201
Cell phone use	190
Overloaded truck	97
Other inattention	73
Sporadic speed	23
Not classified	174

What can we do?

Leon James, author of "Road Rage and Aggressive Driving: Steering Clear of Highway Warfare," recommends three potential fixes for when you feel yourself start to get angry behind the wheel:

Acknowledge the problem.
- Realize you tend to react poorly. When an incident occurs, take a deep breath and give yourself a moment to gain perspective.

Record yourself.
- Set a voice recorder and narrate your drive as if you're doing play-by-play. What you hear during playback may be a little frightening.

Make monkey noises.
- Sing, tell jokes, make weird sounds — whatever it takes to make yourself laugh and lighten the mood.

When Question 6: When are the most dangerous times for women after a divorce (or breakup)? "If a woman is going to be killed by her husband, (legal spouse or just live-in) she is at greatest risk in the first two months after she moves out."

- Dr Reid Meloy, Violence Risk &
Threat Assessment

When Question #7: When will active-mass shooters probably strike? Various criminal psychologists note that 90% of the active shooters had some sort of a triggering event within hours or days of their murders. (This is not uncommon for other crimes.) It's one piece of a puzzle people might alert too beforehand.

When Summary:
> *Wrong place, wrong time.*
> *Wrong place, right time.*
> *Right place, wrong time.*
> *Right place, right time.*

Sure, there are events that led up to things. But I do believe in coincidence, unlike some people. Think of a traffic accident and the quirky events that occur to create one. Think of a crime in the same way. Coincidences even happen inside the best laid plans.

When? Before. During. After. Never stop inventing, asking and trying to answer the when question in confrontations (and life).

The When Question Review.

Who are you within this when category?
What are you within this when category?
Where are you within this when category?
When are you within this when category?
How are you within this when category?
Why are you within this when category?

HOW?

Chapter 7:
The How Questions and Confrontations

In many ways, I find the HOW questions the most interesting and informative. One of the biggest planning, forecasting aspects is trying to figure out just how you will be interviewed or ambushed. How will you react. How will you prepare? Keep in mind there are so many possible HOW questions. Begin the thought experiment...

HOW Question 1: How will they approach you? This answer will cover the pre-assault and the pre-crime lists. One list for anger, bullying, impulsiveness and drunks. The other list is for criminals planning ambush attacks of some sort.

The pre-assault list. This information was first taught to me in the 1970s at both the Army military police academy and Texas police academy. I started this collection then and I've added some. This list is in my teaching outlines since the late 1980s. What are all the things that happen before physical contact from sniper range to stand-offs.

Now, I do not want you to over-emphasize this list as some kind of cure. Just read over the list and keep them in mind. The list was created and repeated here because these tips/events have happened. I have seen many of them when dealing with people for decades in this upset and angry, drugged or drunk "people business" called police work.

When a person becomes stressed, angry and aggressive, his or her body might react, not always, but sometimes it demonstrates some changes. Here are some of these changes that research, history and experience may induce a sudden attack/leap upon you.

Many people suggest that in a real fight situation, a person has no time to read these clues. Sometimes, yes, I agree. But this is not always true. Sometimes there are confrontations and people do have the time to see these tip-offs. Professionals and citizens need to read this list and at least become aware of these points.

Obviously, the clues vary from situation to situation and person to person. But it is better to review the cues on the list, than not, or to ignore that they even exist. I have seen many of them unfold myself on police calls and making arrests.

- His eyes bulge, or he squints.
- He suddenly seems to ignore you.

- He has that 1,000-yard stare. (It use to be a 100-yard stare when I started in the 1970s. I guess what? Inflation?)
- He assesses your body parts and gear as potential targets.
- His mouth becomes dry, creating odd lip and jaw movements. His teeth clench.
- His voice changes.
- He actually, clearly voices violent intentions.
- His words become spastic and distracted.
- He twitches.
- His nostrils flare.
- His breathing increases.
- He takes one big sudden breath.
- His face color changes, maybe reddens or pales.
- His veins bulge.
- His chin tightens, or drops.
- His neck tightens.
- His jaw juts out (dumb but he still does it).
- He babbles as though his thoughts are not guiding his voice. He doesn't babble and actually vocalizes his plans of attack.
- He actually tells you his plans! "Why I'm gonna' hit you…"
- His arms swing, maybe with body turns (a big deal and easy cover for a sucker attack).
- His fingers and fists clench (itchy as blood leaves
the extremities.)
- His fingers drum surface tops.
- His hands shake.

- He suddenly extends a hand to shake yours. Sucker punch?
- Hands go to weapon carry sites on the body.
- He turns away (critical sucker punch set-up).
- His hands and arms travel to near obvious pre-fight postures and positions. He positions his hands high on his chest, neck, chin or head.
- Raises up to seemingly innocent, high positions as in a fake head scratch, like a yawn or a stretch.
- He strikes a pre-fight posture, such as a boxer.
- He rises from a seated position.
- He tries to wander. Verus you or groups, taking offense-defense flanking positions.
- He bends slightly at the knees. (A sporty-like body crouch is never a good sign. I want to say in my experience that I have found one of the biggest tip-offs to trouble is a crouch! Bending at the knees. When the other person crouches. This is a springboard to athleticism. Not only might they attack you, or run off, but in the mixed weapon world we live in, people also have a tendency to crouch and draw knives and guns.)
- He gets too close.
- His body blades away from you.
- He suddenly takes off his shirt, jacket or watch.(Again, dumb, but common.)
- He "expands" his chest.
- Heel and toe tapping.
- Positioning near potential improvised weapons.

- Shirt lift from his belt line (this suggests shirt clearance for a beltline weapon draw).
- Keep adding to this list.

<u>The pre-crime list and info.</u> I am not sure that the average Joe and Joan Citizen grasp the fact that this thrilling, pre-fight indicator list can be quite different than the pre-crime indicator list. The pre-assault list is more for apparent, short-term or long-term, cooking up-heating up personal emotional confrontations. What about pre-crime, and not an emotional, physical, anger fight? What of pre-crime indicators?

Planned criminals might not display any of the aforementioned, pre-assault signs. They can ambush you with a smile, act and approach with a trick, gimmick or question, or jump you from behind. Whether it's a mugging, a kidnapping or long-term con, there's an old enforcement and bodyguard adage - "ACE," which stands for:

- 1: A for Access.
- 2: C for Concealment.
- 3: E for Escape.

1. Access. The criminal has to get to you to attack or con you. There has to be some method of access. Outright approach or hidden in geography, he has got to get close. Even a phone-computer-email-internet scam requires an approach.
2. Concealment. He has to conceal himself for the ambush or conceal the real "him," during the outright approach or con.

3. Escape. He must have an escape plan, a getaway plan. Sometimes that might include murder. Or a suicide by cop, or suicide in general. I am a collector of books on interviews with criminals caught and they are all quizzed about their getaway. As a result, I questioned hundreds of criminals in my cases about their escape plans.

When we caught robbers or burglars in a building, in the act, my first threatening question was, "Where's your car?" He might be working alone or have a driver in a getaway car cruising or parked nearby. If you recall earlier in the WHERE chapter, criminals like crimes near the highways for fast escapes, making all highway locations a bit more dangerous.

Now consider how the ACE acronym might also be planned for a military mission. Access to the mission target, a trench, a border, a fort, camp, city, state, country. Concealment up to and including as much of the target as possible. Escape if the hit and run mission is a success. If it fails, if occupation is impossible, all leaders should have an escape plan.

I believe that while many virgin schools and virgin seminar attendees are so, so happy to hear about all the "fist clenching" and "1,000-yard stares," that the presenter and attendees miss the vital, sneaky crime prevention aspects.

Years ago, Gavin de Becker wrote that entertaining book, *The Gift of Fear.* Some people do remember this book. Why? The stories. Great stories. First editions suggested more of an ESP-ish, Spidey-Sense as the natural fear gift. Neuroscience developments in the

2000s proved otherwise, stating that it wasn't such inert magic, but rather as we stated earlier, as Gary Klein discovered, we are trained to worry, to recognize and react to suspicions and fear largely from learned behavior. Educating the brain. Your "gut" instinct is almost completely a trained mind from previous, vast sources. Almost.

One of my favorite researchers, specializing in stress decision making is Gary Klein, In just one of his great books, *The Power of Intuition*, Gary recalls:

> "For example, I have studied decision makers from military, police and firefighting backgrounds who believed that they had ESP. After showing them evidence, the decision makers themselves admitted that their intuition did not depend on ESP."

The *Gift* stories were indeed thrilling, but take out the cool stories? And what's really left? The skeleton of common advice. Strip out the memorable tales and you have yet another boring crime prevention handout from your local police department. "Lock your doors." "Put up lights." "Watch out for strangers." "Watch out for dark places." Etc.

These pre-crime tips can be as varied as a criminal's imagination. Their plans might include sudden come-from-nowhere ambushes, with approaching-closing-in, acting skills. And these performances may be anywhere, from doorways, sidewalks and parking lot, in and out of buildings, con men contacts. They extend all the way to sit-down, financial, Bernie Madoff schemes. It can happen to you, says author Kathy Scott:

"How do unsuspecting people get duped to begin with? After all, even the most rational people have been susceptible to crimes of trickery. That's because con artists often prey on people's trust and their propensity for believing what they wish was true, especially with get-rich-quick schemes and individual's desire for a quick buck. They let their guard down and buy into what con artists feed them, all in the belief of the scammer and a high rate of return in exchange for a small investment, albeit a shady deal. But the convincing scammer skews the victim into thinking the payoff will come true and the scheme is legitimate.

Each of these con artists have one thing in common: the power of persuasion to swindle their victims. The successful ones exhibit three similar characteristics, psychopathy, narcissism and Machiavellianism (a trait characterized by manipulative and deceitful l behavior and emerged as the strongest predictor of repeated criminal activity.

This trait, along with psychopathy, narcissism, and sadism, could help explain why some individuals are more likely to reoffend. Which have been referred to by Psychologists as "dark" personality traits.)

Those characteristics allow con artists to swindle people out of their money without feeling any remorse or guilt.

Another thing most chiselers have in common are their egos. These extortion salespeople boost the psyche of the perpetrators and make them feel even more confident, thus the description of the con has been termed as a confidence game."

- Cathy Scott, The Crime Book

Each year as a detective, our squad handled these incredibly insane-inane cons that the elderly fell for. Heartbreaking financial losses. The victims would run to the bank and remove sometimes as much as their life savings, just over a phone call con. Thankfully bank clerks and bankers finally would contact us when these old folks showed up at the bank and demand huge amounts or all of their money. And we would rush to the bank. The victims usually refused to cooperate with us when we got there, because the conman speeches told them to remain silent. The conman pitch-deal was THAT good!

The conmen obviously obtained the addresses of the elderly as targets via marketing companies and back then just used the phone books (remember them?) to call old-sounding first names.

Today, we are inundated with the most stupid cons via emails, often places like Africa. You read them, laugh and ask, "who could possibly fall for this!" But, the messages are *purposely* planned to be that stupid, to cultivate the most culpable as victims.

"Street" crimes, outside of phones and emails. Everyone should know about the common, get-too-close-to-you, trick questions criminals use to get mentally and physically close to you. Here are the usuals:

- Asking directions.
- Asking for help of some kind.
- Asking for the time (maybe an oldie but goodie? You look down at your watch and boom!)

- Asking for a cigarette or a light for one. (Another oldie but goodie?)
- Anything-something similar to the above.

Where do you get an overall conman protection list? As with all pre-crime lists -

- 1: Gather intelligence from non-fiction books, (even possibly fiction books. We learn so much from stories.)
- 2: Your local and regional news.
- 3: True crime TV shows. Listen and learn from victims and smart people.
- 4: Anywhere victim stories are told.

So hey, let's make crime prevention interesting again! I mean, doesn't "Pre-Crime" sound cooler than "Crime Prevention?" We can do this. Keep your "scene" just a "scene" and not a crime scene.

"Don't turn your 'scene' into a 'crime scene,' baby."- *Kojack*

HOW Question 2: How long before your perishable skills perish? We humans have always known that continuous training is necessary to stay sharp, yet over time, our skills inevitably degrade. Every area of performance, in every field, experiences this slow decline.

The concept of "perishable skills" has existed since the dawn of humanity. Cavemen practiced spear throwing and tracking wounded prey until their abilities diminished.

Life today is far more complex than merely hurling pointed sticks. Nearly every job, task, hobby, and skill consist of multiple layers of mental and physical performance that must be maintained and refined before they begin to erode.

The first time I officially heard the term "perishable skills" was in police training in the 1970s—yes, that long ago. But even before that, we often heard phrases like "use it or lose it" as common wisdom.

Declining abilities affect every aspect of life, tennis, typing, rodeo, sex, baseball, racing, everything. In many fields, such as law enforcement and the military, there's rarely enough time, funding, or personnel to sustain rigorous training cycles and counteract the inevitable deterioration of skills.

How long before a skill perishes? In pursuit of training efficiency, U.S. state police, the military, and even the business world have categorized skill degradation into three timelines:

1. Perishable skills – Half-life of less than two and a half years.
2. Semi-durable skills – Half-life of two and a half to seven and a half years.

3. Durable skills – Half-life of more than seven and a half years

But how were these timeframes determined? By whom? For what purpose? Organizations need frameworks to justify skill maintenance schedules, but the reality is that everyone's learning and retention abilities vary.

If the path to expertise is unique for each person, then logically, the rate of skill degradation must also vary. The idea of "hours-to-perish" is highly situational and depends on the individual and the specific skill in question. After all, even the widely popularized "10,000 hours to expertise" rule, promoted by Malcolm Gladwell's *Outliers*, was later challenged by numerous experts who pointed out that skill acquisition and mastery differ greatly among individuals. (So please, stop quoting the 10,000-hour rule!)

A lifelong commitment to a craft extends its longevity. Consider the example of professional tennis. Take legends like Serena Williams or Roger Federer. As they age, they inevitably slow down despite constant training. Eventually, they retire, but they remain involved in the sport as coaches or club professionals. Even at 60 or 70, it's hard to imagine that Serena or Federer wouldn't still dominate most amateur players. Though aspects of their pro-game may decline, their accumulated expertise ensures they remain highly skilled long after their prime.

This brings us to muscle memory. The term irks a small group of technical purists who argue that "muscles have no brains, therefore no memory." But it remains a useful and widely understood phrase. Rather than turn this discussion into a biology lesson, let's refer to a succinct definition:

> "When a movement is repeated over time, the brain creates a long-term muscle memory for that task, eventually allowing it to be performed with little to no conscious effort. This process decreases the need for attention and creates maximum efficiency within the motor and memory systems." - "Professor Google"

This fundamental principle underpins skill retention and degradation. The philosopher René Descartes famously said, "Cogito, ergo sum," (I think, therefore I am.) As humans, we will eventually cease to think, and along with that, our skills will fade. I perish, therefore, perishability is inevitable.

But let's not be too fatalistic. As long as we are alive and capable, we should embrace the spirit of our ancestors and continue to train, refine our abilities, and pursue our goals. By consistently honing our skills, we

can remain sharp for as long as possible. And who knows? Maybe, even in old age, we'll still be formidable competitors at the "Caveman Spear Pro Shop and Country Club." You still might end up a pretty good ol' pro at the old Caveman, Spear Pro Shop and Country Club.

HOW Question 3: Response Time 1: "Cold!" How fast is my reaction time? How do I reduce my response times? How do you respond to an interview, argument or ambush in crime or war? How quickly do you start fighting back? This topic will take the next 3 HOW questions. Elements involved in three questions are:

- What's a second?
- Task saturation.
- Zero to sixty scaling.
- Alertness and awareness and the impact of startling.
- Revved up mentally and/or physically.
- Hicks Law versus George Miller's Chunking Law.
- Fighting-shooting cold.
- The OODA Loop "thing."

What's a second is involved here. We hear it takes a "second," or a "second or two" expressions. What does that mean? So, we'll define seconds and milliseconds. There are 1,000 milliseconds in a second. Wow. So, if something took 500 milliseconds to decide, or half a second, how bad is that, really? A quick reference for you - a typical human eye blink lasts between 100 and 150 milliseconds. Milliseconds in passing are hard for me to grasp, even a second alone is hard for me to quantify. Still people like to argue about splitting the hairs of time. The response time argument is about making confrontational decision choices in:

- "half a second,"
- "a second,

- "a second or two,"
- "on up to flat-out freezing."
- The milliseconds and seconds add up and…you're dead, so they tell us. Learn 5,10 or 20 ways to do something and your reaction time is so befuddled and half-frozen in a selection trap called in modernity - "Task Saturation." Task saturation is a state of mental overload where a person has more demands (tasks, information, stimuli) than their brain can effectively process, leading to reduced performance, increased errors, stress, and impaired decision-making, often seen in high-stakes fields. (Let not your heart be troubled, solutions are just ahead.)

Zero to Sixty. Fighting cold, responding cold, isn't about being mugged in Alaska or just a concern for the U.S. Army's 10th Mountain Division. It's about being ambushed in conversation or attack, about starting at body-zero temperature. Zero MPH. "Zero to Sixty" The phrase "zero to sixty" measures how fast a vehicle accelerates from 0 to 60 miles per hour. The faster, the better sports car performance. But in a fight, whether physical or verbal, your absolute zero temperature is the perfect moment to be ambushed.

What does it mean in the bigger picture? How does responding cold apply to everyday life, crime, and war?

Alertness and awareness are the first and most reliable layers of personal defense. When a person is mentally present, scanning their environment, recognizing anomalies, and anticipating potential threats, the body is far less likely to be overwhelmed by surprise, or zero degrees colds. The problem is that most assaults, ambushes, and accidents exploit lapses in awareness.

A sudden stimulus like a loud noise, rapid movement, or unexpected contact can trigger the startle reflex, an involuntary neurological response designed to protect the body. (Hands will usually fly up versus threats to the eyes or neck, fly down versus low threats.) This reflex causes a flinch, or contraction of the muscles, often accompanied by a spike in heart rate and stress hormones. While natural and universal, this reaction can momentarily disrupt decision-making, balance, and coordination, exactly what an attacker or dangerous situation capitalizes on.

Training and heightened awareness do not eliminate the startle reflex, ***but they shorten and control it***. Alert individuals are less easily startled because their brain has already assigned meaning to what it is seeing and hearing. When the unexpected does occur, trained awareness allows the startle response to rapidly convert into purposeful movement, protective positioning, evasive action, or immediate counteraction, rather than paralysis. In short, awareness buys time, and time buys options. The goal is not to suppress the reflex, but to ride it, turning a split-second of shock into a functional response instead of a catastrophic pause.

The startle reflex has gone down some rabbit holes in martial training, dumbing things down with overdoing the topic. For example, there is a gun instructor who demanded that before EVERY pistol quick draw, one must first throw their hands up in a mandatory startle, then down to grab the pistol

. Really bad muscle memory. Another instructor tries to overuse the raised bent arm startle into solutions that are not the best solution to various problems but still adhere to his marketing program. This is bastardized self-defense for marketing, brand motives.

BOO! Above and simply the best book ever on the startle is this science textbook. It is expensive but it is a must-have, must-read to anyone who wants to truly understand the subject and destroy the marketing myths that abound in the martial world.

Hick's Law. It all starts back in the 1950's with "Hick's Law." It declared that all mental and physical decision-making takes "time." This law came from a simple, two panel, flashing color light test. The testee sits, sees a one of a few color lights flash on. He reacted by pressing the same color button on another panel. One decision. One movement. Not a two or more decision-making test, just one. But with this solo test, Hick (and his associate Hyman) declared:

> "...that the time it takes to make a decision increases logarithmically as the number of available choices increases, meaning more options lead to longer decision times and

potential confusion or "choice paralysis" for users." And Hick's Law was cut in stone. Hick and Hyman ended their study with no solutions offered. Leaving us astray, slow and dumb.

Back in the 1990s and 2000s, numerous martial arts sales-marketing-mythologies were in play through MA magazines. One was shoving Hicks Law down our throats, (oh, another was the heart rate chart from the old police PPCT course and well, don't get me started here on that…) I am quite sure that Hick and Hyman, back in the 1950s, ever dreamed that their hyper-jump from a "one-button-test" led to estimate times for multiple selections and would be used for combatives and…and everything in life, when you think about.

The martial arts conclusion back then was that you must become a one-trick-pony, when you might be, could be a three-trick stallion. In other words, "dumbing you down." Face it, two or more decisions and you're just a slow-calculating slug.

Solutions and better understanding. Newer, better studies do have solutions, and that's why newer studies are so important to find. But some studies aren't so new on this subject at all! And one big solution is what the experts call "chunking." When did chunking arrive "in the field?" Ironically, also in the 1950s! From one scientist, George Miller and his "chunking" discoveries. Let's look at A.I. for this history...

"When Chunking Became a "Good Solution." Chunking, grouping multiple small actions or pieces of information into a single larger unit, became recognized as a powerful solution for improving human performance beginning in the 1950s, with its scientific foundation solidified in the late 1950s and early 1960s. This George Miller book is often cited as the formal birth of "chunking" in cognitive science with:

In 1956, George A. Miller wrote "*The Breakthrough. The Magical Number Seven, Plus or Minus Two'*. In it, Miller demonstrated that the human brain has a very limited working memory capacity. People manage this limit every day by grouping information into chunks, which allows complex actions or data to be processed as a single unit. This became the foundational research proving chunking is not just helpful it's biologically necessary."

While Hick's Law tells us that more choices slow you down and every extra decision point adds precious time to your response. Chunking is the antidote. Chunking is the brain's ability to make-take multiple separate choices and weld them into one single, automatic action. Instead of three choices, you have one. Instead of three decisions, you have zero decision makings. For example, in untrained people a block, strike, and step might be three separate tasks, each requiring its own mental confirmation. But trained fighters merge these into one smooth "burst," the brain fires off the whole package like a single shot.

1. Block and strike (two steps).
2. Block-strike simultaneously (one step, a chunk.)
3. Block strike and then kick (3 steps, but can be 1 step when practiced as 1 chunked series).
4. Have a few series of more 2-3-4 steps.

5. The martial I.Q. of the instructor builds the smartest chunks, based on the reactions of the opponent. For one example, just about everyone does jab, cross, hook. BUT! Your jab and cross knocks-keeps an opponent out of your hook punch range. Not a smart chunk.
6. Everyone is different by age, strength, size, mental acuity, and potential.

This is the same chunking process musicians use to play a song without thinking of every note, or how drivers shift gears, brake, and turn without mentally verbalizing each step. The brain stores the group as one high-speed motor program. Once chunked, the action bypasses slower, deliberate thinking and moves into the fast lane: reflex-level, sub-cortical, near-instant. Even in our everyday lives, we chunk. If we mow the lawn, or take up the piano, cook a meal, we immediately start chunking. We chunk so much, which Law should make a decision-making priority lesson? (By the way, reps certainly count, but the 10,000 reps "rule" has been shattered by research. People are different and need fewer or more reps, usually less. You can easily look that up. Drop that one too.)

In self-defense, in confrontations, chunking turns a longer sequence into a single solution. You are not deciding between six blocks or nine strikes or contemplating salt versus pepper in the kitchen. You already know the recipe. You are executing a built-in "package response" you've done many times. This dramatically shrinks Hick's Law importance. Instead of being bogged down by options, you've compressed choices into one immediate tactic.

The bottom line is simple: The more you chunk, the less you think, and the faster you fight. "Chunking turns choice-complexity into simplicity, and simplicity into speed—the only real cure for the hesitation Hick's Law warns us about." (And that's an A.I. quote again, not me). Your job is to investigate a lot of things, experiment, and then get your favorites down to suit you. Fit them into chunks as much as possible. A.I. continues –

"In summary, Miller's concept of chunking is a powerful mechanism for organizing information that makes it easier to process, thereby allowing individuals to manage a larger overall number of options while keeping the immediate number of choices low, thus resulting in faster decision times than Hick's Law's."

Translation. There are three teaching "moments here…

1. Bad. Counter-productive. If you teach 20 ways to handle a hook punch and make people memorize all of them, this is a form of task saturation.
2. Positive. Productive. If you teach-show twenty ways to handle a hook punch and then order a practitioner to experiment with the 20 and then order them to memorize only the 1 or 2 best-suited-for-them (via chunking), that is NOT task saturation. THAT is a good training doctrine. Teach them, inspire them with what chunking means.
3. Bad, Counter-productive. Yet, if you show-teach these 20 ways and tell them nothing

about a personal selection process, that is incomplete teaching and incomplete system doctrine.

Rather than selecting choices from a series of hand strikes, in the methods of chunking and the principles within *Conceptual Learning,* the boxer does not waste a second selecting specific punches but rather picks one overall decision-concept like "punch many times," or picks sets of pre-trained punches melted into one movement. The trained body then takes over, following paths learned from prior repetition training. Another example is the generic, targeting message "Hole! Hit into it!" And foregoes the rolodex of selecting a punch...selecting another punch...selecting another punch.

Hicks versus Miller? Hicks no solution. Miller, solution. The irony to me is...both Hicks and Miller were produced in the 1950s. Miller is smarter and deeper, inspiring, overcoming, and more real, especially for martial choosing. Yet I presume academia advertised Hicks more? Or didn't know about Miller?

Instead of throwing the name "Hick's Law" around, maybe we should throw "Miller's Chunking Law" around. It's better, smarter. More real. Yes, I get it, Hicks law existed in the "pantheon" but is just a steppingstone to Miller. Miller is way more important to training.

By the way, there are other studies that leave Hicks in the dust. Pick ones like Miller as your teaching-training doctrine. Let's chunk Hicks law.

Question 4: Response time 2: The concept of shooting cold.

You have just read the prior essay and all that counts again here. But wait! There's more!

A cold barrel means a cold shooter. Cold shooting competitions are challenging and realistic, revealing how shooters perform without warm-ups. I became interested in fighting-shooting cold through military police shooting training in the early 1970s. Gun experts use the term "shooting cold," often in reference to snipers or hunters those who must take a precise shot under pressure with no warm-up. In some competitions, like a

> "one-shot" contest, participants fire a single, cold-bore (clean, unfired rifle) shot from 1,000 yards. Time and accuracy determine the winner."

The mental challenge of fighting cold. Who wants to shoot, fight or debate argument points ambush-cold, face any confrontation ambush-cold? Most of us prefer a little warm-up. In citizen, police and military training and shooting competitions, participants typically get briefings, walk-throughs, or even dry-runs before the real course begins. They might even "test" their guns first. After all, jumping into complex drills cold could be unsafe. These are all forms of warm-ups.

Interestingly, some of the best shooters I know struggle when shooting cold. That's why serious shooters track their first cold shots over time. Many find with their cold-shot records that, despite their skill and dedication, their cold shooting doesn't perform much better than those who practice less often. Experts admit that shooting and fighting cold is difficult. It involves situational awareness, mental and physical and "revving up."

Revving up. subliminal preparation. Even when aiming for a cold start, mental and physical preparation plays a role. A shooter may arrive at the range, set up gear, sip coffee, chat with range officers, and listen to instructions, all of which gradually warm up their mindset. They are not stepping into a completely frozen, ambush-like situation. The same principle applies to fitness. If you pack your gym bag, drive to the gym, park, and walk inside, your body and mind recognize the routine. You are subconsciously preparing, warming up before the workout even starts.

I once saw a man in a gym parking lot pause, remove his cap, and whisper a prayer before entering. Was he psyching himself up? Asking for strength? Protection? He was revving up. Standing still. Praying.

Chilly. The power of mental preparation. Sometimes, preparation isn't always short-term. We've all attended political debates on Thanksgiving, or a shooting class, a martial arts tournament, or competition that we anticipated for days, even weeks. The inner engine revs long before the event itself. Mental preparation is powerful. Studies suggest that visualization alone can enhance performance.

I remember hearing martial artist gossip in the early 1970s when I started in Kenpo, talk about systems that had its students train sometimes in dark rooms, imagining their moves over and over. At the time, it seemed ridiculous, but research now suggests it works.

Even mundane routines can serve as mental preparation. A retired British SAS veteran once told me that when he irons his suit before work, he begins mentally preparing for the day with each stroke. Whether you're a security guard, police officer, lawyer, truck driver, or accountant, simply getting dressed for work starts priming-revving your mind.

<u>Your cold handgun in your cold holster</u>. Handgun training often fixates on the speed of a pistol quick draw, drawing and shooting in split seconds. I wish that the worlds of hand, stick, and knife combat would train with the same level of deep concern that the firearm world has for topics like drawing under stress. I do but most don't. Most just train with the weapons magically in their hands, forgetting that they have to first draw those bloodsuckers out.

Despite the many variables involved in a gunfight, extensive time and effort are dedicated to the prepared, alert person's ability to draw and shoot at a bullseye.

Every pistol course includes some form of start-stop, split-second timer to measure draw speed. But drawing is more complicated than simply standing on a range with itchy fingers, trying to beat a beeping timer. It's necessary, but far from perfect.

College institutes like Force Science and various sports labs test individual draw times under different distractions and conditions. This is crucial. Every pistol course should emphasize that a gunfight draw depends on:

- The situation.
- The exact cue-reason to draw and shoot a human.
- Your reaction time to that cue-reason.
- The type of gun you have.
- The type and placement of your holster.
- What clothing are you wearing.
- A note, most criminals don't even use holsters.
- Another note, not all gunfights start in an old West-style quick-draw standoff, yet disproportionate time is spent training for and with this rare "showdown" scenario. Working through these with simulated ammo for decades now, I will report that very often both shooters wind up shot and/or "dead."

Each of these factors can add critical seconds to your stress-laden survival draw. A general summary of draw times comes from range shooters, those prepped, dressed for and revved to draw and shoot.

- Beginners might take 2–3 seconds to draw and shoot, under non-ambush conditions. Reality distractions make things worse.
- The average skilled shooter takes about 1–2 seconds. (From concealed?) Reality distractions make things worse.
- Elite competitive shooters can draw and fire in under half a second—again, under controlled conditions. (From a concealed carry? A lot of timed range shooters run their guns from an open carry, not real, everyday clothing.)

The myth of the western-style showdown shootout. When we organize these showdowns with sims ammo, often both shooters are shot. Shot dead. Range, bullseye shooting, which you must do, innocently winds up in this structure with differing instructors' invention and re-invention and tricks, yet all still doing mostly the same thing.

Drawing and shooting at bullseyes. Which you have to do, is still very incomplete for combat survival. Force Science refers to this as "internal focus," getting the draw and the aiming down, while not being shot back at. External focus is worrying about being shot at and the who, what, where, when, how and why of the enemy. *THAT* is my mission to cover in my *Survival Centric: Gun* course.

<u>*Frigid cold? How about asleep?*</u> Now, consider an even colder situation. Being asleep and then shooting. Some studies have tested police officers' response times upon waking. In one experiment, state troopers were tucked in to sleep in a controlled environment, told that when

abruptly awakened, grab a bedside gun and shoot a target at the foot of their bed.

The results? Not great. Often, they were downright poor. They were sound asleep! A similar study allowed officers to wake up naturally in the morning and recall on their own the assigned task of immediately shooting upon waking. They performed slightly better but were still groggy and slow. Where does this information fit in our "cold" and "frigid" considerations? Frankly, I'm not completely sure, it speaks to prior-awareness-alertness-revving up, but it's fascinating.

The moment you start engaging in physical activity, jogging, training at the gym, stepping onto the range, or even chatting with an instructor, your internal gears begin turning. The first round of cold shooting might not be as frigid as you expect. But an ambush? That's like a zero-to-sixty situation, like being half-asleep? And that's a different kind of cold.

Life Is either an interview or an ambush. This phrase remains true. The greatest armies, soldiers, cops, and fighters have all fallen victim to ambush, the raw power of surprise. And yet, there's always an armchair expert screaming, "They weren't alert!" or offering the sage advice, "You must always stay alert."

We often hear, "You don't pick the time and place of your attack; the enemy does." For the everyday person, or someone on routine patrol, an attack will likely come cold. If you're a citizen, you may suddenly find yourself in a crisis, forced to think cold. Training for "cold fighting" and "cold thinking" is crucial.

One of my veteran and gun-guy friends, after reading this when it was first published in 2011, said:

> *"Hock is right about this. I suck at shooting cold, but that's how I'll have to shoot, cold, stepping out of a Waffle House and suddenly in trouble on any given night."*

Cold or hot crime? One last point on the cold-hot spectrum: Criminologists and neuroscientists analyze crimes by categorizing them as "hot" (emotionally driven) or "cold" (calculated and detached). Understanding this distinction is another piece of the puzzle in how violence unfolds, and how we train to respond.

How Question 4: Response 3: The "OODA Loop thing and how then do I reduce my response times? Well, you just read HOW questions 1 and 2. Chunking moves shorten response times. The next conundrum is the OODA Loop. The loop exists within the startle-surprise world. A good fake creates the stall-startle.

At the heart of the split-second decision making, counter ambush is "ye old" so often quoted OODA Loop presentation. *Observe, Orient, Decide, Act*. It is name-dropped by many people wanting to sound inner-sanctum-secret, military educated. It would be hard to discuss sudden, pressurized decision-reaction without gracing the pop, OODA subject, regurgitated by so many. "General Wikipedia" defines it as:

> "The OODA loop is a decision-making model developed by United States Air Force Colonel John Boyd. He applied the concept to the combat operations process. It is often applied to

understand commercial operations and learning processes."

We all know that the element of surprise and reaction can be as big as a world war invasion and as small and simple as a foot fake or head fake in football, rugby, soccer and kickboxing. The reality is actually quite simple:

"Fake-surprise (boo!) versus reaction time."

Bingo. Simple. That simple. Back to the *Boo* book again. I don't think we need a chart the size of a doorway, like the latest OODA Loop, demo diagrams that have ridiculously evolved to explain this simple, oh so simple, "boo/surprise" idea.

Having read the interesting biography on Boyd, when he did his Air Force presentations on his Loop, some attendees thought it genius, others yawned and asked, "why are we here." I lean more to the yawn category. Techno terms are used like "stepping into stalled decision-making brain process."

Subsequent charts were drawn that resemble the architectural plans of the Pentagon building. But OODA is as simple as a foot fake in football or soccer. You are just making a surprise move within any endeavor. Oh, but the military jargon, get this militant review-

"The real OODA Loop, is a nonlinear, adaptive system integrating information and energy as currencies that flow through an organism or organization."

(Oh my, the ostentatious, dense military jargon.)

Big ambushes and small ambushes. You know that well before Boyd, fighters and the militaries were faking out opponents with plans. They were using big battlefield strategy fakes on down to spear fakes, sword fakes, knife fakes, and hand to hand combat fakes. The element of big and small surprises has defeated the greatest militaries of the world.

Reducing response times; A three-part HOW question summary. Try what the military calls: Immediate Action (I.A.) Drills. I.A. drills are George Miller Chunking. I first learned about all this Ambush/Counter-Ambush in the U.S. Army in 1973. This was not new. They did not use the word-concept of chunking, nor OODA, yet, there it was. And it was a big deal. They trained us in what was called back then I.A. things done so many times that you may well jump right into that response groove when ambushed.

It is reinforced by many, many chunked repetitions. Here are some of my old U.S. Army manual ideas on the ambush drill idea that relates to citizens and police.

"Immediate action drills are drills designed to provide swift and positive reaction. They are simple courses of action, done immediately. It is not feasible to attempt to design an immediate action drill to cover every possible situation. It is better to know few immediate action drills for a limited number of situations that usually occur in a 'combat' (and we add crime to this) area. It is all situational, probabilities versus possibilities."

Here are some proven methods that improve overall reaction time and performance:
- Sequential Learning, the stringing of tasks working together like connected notes in music really reduces reaction and selection time. Several things become one thing.
- Conceptual Learning, is another speed track. In relation to survival training, this means a person first makes an either/or conceptual decision like "Shoot/Don't Shoot" or "Move In/Move Back."
- Implicit and Procedural Memory, in Dr. Lee Dye's 2009 article for ABC News, *How the Brain Makes Quick Decisions,* he reports, "People have been helped by a kind of human memory that scientists have been struggling to understand. People use "implicit" memory, a short-term memory that people are not consciously aware they are using.

    Doctors Ken Paller at Northwestern University in Evanston, Illinois, and Joel L. Voss from the Beckman Institute and the University of Illinois at Urbana-Champaign have conducted long-term research on this subject; and while they did not specifically involve athletics, the conclusions are consistent with other researchers who are also studying how top athletes can make split-second decisions and take action. How does a batter hit a fastball when he has to start swinging the bat before the ball even leaves the pitcher's hand? "He relies on visual cues, even if he doesn't know it." Athletes and people (even driving cars) learn to predict and act and react spontaneously based on very little information. One way Is implicit memory.

- Implicit memory. (IM) is a type of memory in which previous experiences aid in the performance of a task without conscious awareness of these previous experiences. People rely on implicit memory in a form called procedural memory, the type of memory that allows people to remember how to tie their shoes or ride a bicycle without consciously thinking about it. Implicit memory taps into procedural memory.
- Procedural Memory. (PM) One more related subject in this chain of memory and performance. PM is connecting small multi-tasks and problem solving. Examples of procedural learning are learning to ride a bike, learning to touch-type, learning to play a musical instrument, learning to swim, and performing athletic tasks like sports. This includes martial moves, fighting, self-defense, combatives, whatever you wish to call it. Experts report that procedural memory can be very durable, however perishable, like any

task. And the physical fitness to perform these tasks may not be so durable.

- Crisis rehearsal. In police training decades ago, the instructors told us to "crisis rehearse." The message was, "There is a lot of down time on patrol. Look at every business in your district and imagine them being robbed. Where would the robbers park? How would they do it? Where do you park in response. Rehearse every crime, every place in your beat." (Yes, the Ws and H!). This was one of the most important, smartest pieces of police advice I'd ever received.
- And finally...Myelin. Can you improve your reaction speeds? By a second? By a few milliseconds? Try it out and see. How long is your martial and sports world second? Myelin is mentioned. "Myelin insulates nerve fibers, allowing electrical impulses to travel quickly and efficiently between neurons. Myelin is made up of protein and a fatty substance that forms a sheath around nerves, including those in the brain and spinal cord. Myelin and repetition training are closely linked, essentially creating 'muscle memory.' The more you repeat an action, the thicker the myelin sheath

becomes, making the skill feel more natural and effortless."

- Medicine Plus
- Repetition turns a brain's dirt road into a brain's superhighway. The construction crew builds with Myelin.

HOW Question 5: How do you control yourself when confronted? We shall take on three big subjects here.
1. Adrenaline.
2. Tactical breathing.
3. Mentality.

1: Control adrenaline? Adrenaline the ogre? Decades ago, an adrenaline burst into your body was sold as some kind of poison, a boogeyman in martial, police and military training. Though I have been researching and writing on this for decades, it would be hard for me to pinpoint when the negative campaign happened. But, some people back then must have discovered, an opportunity to sell martial training from a different marketing angle.

After several introductions of the subject in the 1990s and 2000s, poorly informed "friendly neighborhood ku-raty" instructors began name-dropping things like the aforementioned "OODA Loop," the "Startle Flinch," and the Adrenaline Ogre" to sound all "sciencey," and ever-so-learned. (Don't get me started again on the PPCT course heart rate chart. I refuse to even show it in the book.

Amongst this wannabe crowd, they preached that just every hesitation or false step, most every human error, nearly every problem a person had small or big, whether they were ambushed or not, came as a result of evil adrenaline.

- Ogre Myth 1: It robs your vision down into a cardboard tube. The opponent steps to the side and becomes invisible. Yes, invisible. A very popular instructor makes this statement.
- Ogre Myth 2: Snatches your hearing.
- Ogre Myth 3: Cripples your ability to think, act and perform.
- Ogre Myth 4: Made you a big, slow, numb, gross motor dummy.
- Ogre Myth 5: Pooping and peeing in your pants with each stressful event and automatic at a certain high heart rate we have all surprised and know better. (We can thank PPCT for that one.)

But is adrenaline really such a poison? Look at the definitions and descriptions:

"Adrenaline is used in both technical and nontechnical contexts. It is commonly used in describing the physiological symptoms that occur as part of the body's response to stress, as when someone is in a dangerous, frightening, or highly competitive situation, as well as the feelings of heightened energy, excitement, strength, and alertness associated with those symptoms. In figurative use, it suggests a drug that provides something with a jolt of useful energy and stimulation. - Merriam Webster

"The adrenaline response has a number of very specific effects aimed at maximizing survival, mediated by circulating epinephrine and cortisol." - Braunwald et al

"These effects include a state of heightened alertness, increased energy with which to meet a potentially difficult situation, and augmented muscle strength. In preparation for battle, chemicals are released into the blood to facilitate clotting, and blood vessels in the skin are constricted to prevent heavy blood loss in the event of wounding.

Similarly, blood pressure and heart rate increase and the kidneys retain water, all in support of tissue perfusion and the maintenance of fluid volume in the event of sweating or blood loss.

In addition, the spleen deposits red blood cells into the blood stream in order to increase oxygen delivery to muscles and pupils dilate to let more light into the eyes in order to increase visual acuity." - Dr. William F. Ganong, 2001

Performance problems? Not always from adrenaline. To lump all performance problems into one cause is to do a disservice to training doctrine. Once you recognize this truth, you can treat the real, individual poisons.

- A police officer may not think clearly just because she's worked a double shift.
- A distracted soldier may freeze just because he was cleverly ambushed. A citizen may not put their key in the door of their home fast enough when being stalked, not because of adrenaline, but because they have simply never practiced putting their key in their door very fast, and/or in the dark.
- You may not reload your gun fast enough simply because you haven't practiced doing that task on the ground, sideways and in the mud, as well as fast. It's different.

It is also a "zero-to-60" issue. How dull and unprepared were you, the very few seconds right before you were confronted with a shock or action? Zero-to-60 responses are the toughest. Which is why the ambush works and adrenaline might have nothing to do with it.

About 30%-ish adrenalized may be the best. As a cop, I have always done my best when I have been at a stage or level of being "half-adrenalized," for lack of a better description. But, needless to say the human race is alive today through all its trials and tribulations thanks to adrenaline, not because of it.

2: Beathing and control. I have performed best in my life when I have been slightly or somewhat adrenalized. Some experts might call this time, "riding the flow" or "in the zone," whatever phrases. For me, I think the zone and the flow are really about being in a half-adrenalized state. And I am not talking about marathon running here, something a bit more... dangerous.

Looking back into my past, some of my worst performances as a cop have been when I have lost this overall control, let adrenaline run amok. And let me tell you a good ambush can snap the sense right out of you. A car going zero-to-60 in seconds might become difficult to control on the various roadways of life. Unless it's a race car. Are you a race car driver on the racetrack of life? Some people are. Most not. Most of us need consistent road work. Check the tires!

In my past, mental and physical distracting problems like the lack of sleep, hangovers, family problems, constipation, you name it, have interfered with my job performance in many ways; but these problems also interfered with my ability to handle surprises and to control my temper. And control these adrenaline rushes.

How to hit a somewhat half-adrenalized state? And stay there? Get into that flowing zone? It's a connection to your personal calm. There are tons of training programs, mostly for civilians, and unfortunately with a lot of voodoo, buzz words and terms. I mean ...I get it... yeah...but verbalizing a mental quest for your "center" in a sudden gunfight? You have to first build, find and flip a switch of a prepared process in your brain. And tactical breathing helps.

Human kinetic sciences say that good breathing techniques bring the mind and body together to produce some amazing feats on the sport field. Feats well beyond the subject of simple calming and relaxing mental thought control. All medical and psychological experts agree that there is one common thread used to counter some of the anguish of anger, pain, and fear. Breathing! Yes, simple breath control.

No matter who the experts are, from the toughest, scarred, tattooed war vet to the armchair PhD or robe-wrapped yogi guru, all agree that deep and slower breathing can really help control and stabilize the body under stress. You don't have to seek out a monk in China or contemplate your navel in front of incense or a pink candle. This universal, raw method truly bridges the gap between the police, the military, the martial artist and the citizenry.

In today's mental health industry, stress management is a major challenge as well as a very prosperous business. Meditation is different. It is hard to meditate when someone is punching your face. For that industry, the majority of problems are marital, jobs, rush hour traffic, raising children, and the like. "Civilian-life problems."

We all have sudden and slow-burning stress problems that involve distorting our bodily chemistry. We all have "before, during, and after" stress problems. Citizens in "everyday life," and soldiers and police have different kinds of stress in their everyday lives.

What do all these people feel in their bodies when they get anxious or threatened? Here is, once again and just for the record and this essay, the classic list.

Rapid heartbeat, shallow, rapid breathing. Tense muscles. Physiological changes take place in the body. The brain warns the central nervous system. The adrenal glands produce hormones (adrenaline and noradrenaline). The heart beats faster. Breathing becomes more rapid. Fast breathing. The person's body is getting ready to do one of two (or more) simple things - confrontation or departure."

Back to this very critical term of "fast breathing," because breathing is the key subject of this essay. A normal breathing rate for an adult at rest is 8 to 16 breaths a minute. Most people are not conscious about their breath count or the way they breathe, but generally there are two types of breathing patterns.

> 1. Shallow, rapid thoracic (chest) short breathing.
> 2. Deep or diaphragmatic (abdominal) breathing.
> *Note:* Numerous human biologists say that 4 shallow breathes can count as one deep breath.

The stressed body needs oxygen, and we need to pump oxygen to the performing muscles. Slow-twitch fibers affect muscle endurance provided enough oxygen is delivered to them. Fast-twitch fibers, which affect muscle strength, develop peak tension quickly and fatigue easily. That is one reason why slower nasal breathing, not fast mouth breathing often works better. Nasal breathing runs by the vagal nerve, which sends calming messages to the brain. Breathing through the mouth bypasses a large portion of the nasal cavity process of warming, moisturizing, and eliminating particles from the air before it reaches the respiratory system. Breathing through the mouth also further triggers the fight or flight response! Sort of a double whammy, if you will.

I do like the overall term "Tactical Breathing" used by many, for the before, during, and after. Three parts to it. Three "events." This allows us defined measures for each phase. Combat Breathing should be a sub-category under Tactical Breathing. (Remember, good training programs are all about doctrine, doctrine. Doctrine! Words. The proper skeleton allows for the proper fleshing out.)

Numerous police and military people call this wrestling with breath under stress a "Combat Breathing Event." A singular event? Combat breathing to me should cover an overall bigger "event," as in the "before stress, during stress and while-it's-happening stress categories."

Tactical Breathing (in three-part harmony):
1. Before the event – preparation breathing before the event.
2. During – the combat breathing, hardest to remember to do because you are distracted.
3. After – breathing after the event to recover.

Combat Breathing means breathing WHILE in combat. For many real performance experts, combat breathing is just the "act of doing." Doing what needs doing with what you have on hand.

During the fight! Athletes must learn the laws of pneumatics, the science of pressurized air, in this case, as a power source by absorbing and transmitting energy in a variety of sports situations. Most commonly, we know about the exhale when you say, for example, push up in a bench press. Exhale, if you can (as sometimes you can't) when you punch or strike. Firearm shooters and combat shooters (snipers or otherwise) constantly worry about breathing during their trigger pull, but in the chaos of combat, you have to strike or shoot when you have to shoot. Breathing pace often be damned.

So, deep breathing. The only problem is remembering to do it inside the shock, surprise, fight, ambush. It seems that fast breathing is a dirty trick in the biology of survival, doesn't it? It is so easy to forget to breathe when the knife is dropping onto your face. But still, you must try.

What about breathing before the anticipated fight? Remember, not all fights are ambushes. Here is a trick I learned decades ago from police instructors in the 1970s, and one I continue teaching to emergency response folks. I would suggest connecting this type of breathing every time you turned on your police car, ambulance, fire truck sirens, or answer any "hot call." Hot calls equal calming breaths. When you hear the siren? Start the proper breathing right then.

Stairs! "What good does all this running do when I can't dash up a flight of stairs?" I'd ask myself, at the top of the stairs. Another trick I noticed was no matter what great shape I was in as a younger man, how far and fast I could run in miles, yet often when I dashed up a flight of stairs at the police department or elsewhere, I would still become winded.

At one point in my life, I could run about a 7-minute mile, but a sudden, short dash up the stairs would bother me and my breath! But it is a classic "zero-to-60" situation. I swore then that I would slow/deep breathe every time I climbed any stairwell anywhere. A habit. Every time I looked at a stair step! I made it a personal habit. This turned into a major survival tip as we chase and even fight on stairs frequently. Climb any stairs anywhere? Deep breathe. (And, of course, you could run stairs as a workout, another testimony to practicing exactly what you need to do, reducing the abstract)

But as my wife Sandy will tell you - she participates in marathons and these insane skyscraper-stair climbing events – even then the occasional flight of stairs gets one a bit out of breath. A zero to sixty thing.

Calm zones, flow and "the zone." Also, for many years I ran a local martial arts class. Often, I would have to spar/kickbox every student in the class, weekly. This was demanding, however I discovered within myself a certain, calm zone of performance where I could think, coach, and kickbox everyone rather tirelessly! I "recorded" a calm spot in my physiology. This zone. Whatever you call it. I could often find this very spot under police stress and confrontations, too. In a way some might call this a biofeedback method.

One good descriptions and access of and to the "Zone" and also the "flow," I found in the Dr. Daniel Goleman's bestseller "Emotional Intelligence."

> "…yet the zone, the flow is an experience almost everyone enters from time to time. There are several ways to enter. One is to intentionally focus a sharp attention to the task at hand. The quality of attention is relaxed yet highly focused, not straining. Their brains quiet down."

Otherwise known as "Runner's High," Runner's high is a temporary state of euphoria, reduced pain, mental clarity, and emotional calm that occurs during or after sustained aerobic exercise, most famously long-distance running, but it can also happen with cycling, swimming, rowing, rucking, or fast hiking, or…kickboxing, or randori.

What causes this high? Originally credited to endorphins, research now shows it's driven largely by the same system involved in mood regulation and stress relief. The body systems that control mood regulation and stress relief aren't a single chemical; they're a network of interacting neurochemical systems. The result is often described as:

- Effort suddenly feels easy.
- Pain fades into the background.
- Breathing and movement synchronize.
- Thoughts quiet; awareness sharpens

This aligns closely with what many martial artists, soldiers, and endurance athletes describe as a flow state, something you've likely seen (or felt) in long training evolutions.

How to reliably achieve the zone, the flow, that state? Experts say it's about duration, rhythm, and intensity.

- Typically appears after 20–45 minutes of continuous aerobic-style work. Beginners may need longer; trained athletes sometimes reach it sooner.
- Stay aerobic, not anaerobic.
- Work at a pace where you can breathe through the nose or speak short sentences.
- Roughly 60–75% of max heart rate.
- Too hard? Stress response and no high.
- Settle into discomfort. Don't fight it.
- The high often arrives *after* the initial fatigue and mental resistance.

- Pushing calmly *through* that phase is key.
- Minimize distractions. Let the mind drift or lock into breathing and stride.
- Train consistently. The so-called "high" becomes easier to access over time.
- Why some people never feel it.
    - Exercise sessions too short.
    - Training always too intense (HIIT only).
    - Chronic stress or sleep deprivation.
    - Expectation chasing ("trying" to feel it blocks its way).

<u>*"Combat, tactical" breathing steps.*</u> Responding to emergencies, climbing stairs or crossing the street, the point is to pick a good time to breathe like this and make that practice an engrained habit. Once in "combat," you have a lot going on, and your body wants to immediately breathe a certain way. You make it breathe *your* way. The best way you can. Good instincts. Good training. Good coaching. Good mental tricks. Good luck. A car going zero-to-60 in seconds becomes difficult to control Technically, tactical breathing goes like this. Breathe in through the nose for four or more counts. So proven, from Lamaze to Basra. It works. For the record, U.S. Military training manuals describe this advice and process:

This technique, known as combat or tactical breathing, is an excellent way to reduce your stress and calm down. This breathing strategy has been used by first responders, the military, and athletes to focus, gain control, and manage stress. In addition, it appears to help control worry and nervousness. Relax yourself by taking three to five breaths as described.

- Breathe in counting 1, 2, 3, 4. Deep into the lower lung and the upper "belly" should expand, unlike a shallow breath. Today's human biologists suggest training to fill your lungs and then - take in a last sudden, like "reverse gasp" inward. This sucking in, really fills the lungs.
- Stop and hold your breath counting 1, 2, 3, 4.
- Exhale counting 1, 2, 3, 4, through the nose and mouth.
- Repeat.

Tactical breath and the open mouth. I saw this ad-photo and believe me, I take sub-zero joy in criticizing systems, schools and people. I have removed the logo-names, etc. I get the hotshot marketing, it's action packed and inspiring. Tribal. And many martial arts just love a lot of yelling. What then immediately captured my attention here?

Please keep your mouth closed in a fight. Sorry to use your picture, girl.

The unnecessary open mouth (this one way wide). As a police detective for almost two decades, I had to work so, so MANY broken jaw, serious-bodily-injury cases from simple blows to the jaw in simple, stupid fights. This injury can be quite devastating. Jaws wired shut! About SIX weeks healing. (Explosive breathing does not mandate open jaw-ness.)

There is a reason God made mouthpieces. Mouthpieces while training and competing are fantastic in sports and can also create terrific, 'leftover,' jaw-clinching habits. A jaw should be clinched in any and all fights.

Boxers, kickboxers and all UFC-ers wear mouthpieces in training and develop this classic, "closed-mouth." They don't have to have magic yelling. No. Neither must you, Burst-of-breath-called pneumatics can be done WITH teeth clinched. And you do not have to yell with every strike. (A pneumatic breathing aid is described in which air pressure is used to compress the thorax and abdomen. Tightening the torso can help lessen the pain-injury effects of incoming strikes.)

Keep your mouth closed before, during and after your strike, like a boxer, kickboxer wearing a mouthpiece. Yelling, open mouth bursts are an unnecessary "traditional" risk, because you might THINK you can control your "open mouth" seconds but then...I know it's hard to let go of the traditions. The dogma but let it go! No excuses. Be free. Evolve. End it. (And of course NOTHING against this great girl in the photo. She is just doing her thing, as trained.)

3: Mentality. Pretty sure by now in this book, you know and have some ways to keep your head on straight. For even more ways, keep reading.

In Summary. Tactical breathing is three parts, the before, the during, and the post of fighting with hands, sticks, knives and guns. While there are some similarities to a meditative style of breathing, "tactical" breathing is not for the yoga mat. It is almost impossible to forget to breathe properly in a meditation class. It's hard when you are chasing a car at 100 miles per hour as a cop or fighting someone while sliding down a muddy hill in the rain. (Both and more happened to me so I thought I would mention them.) Controlling adrenaline? Training and experience will help, but that's not all.

> "Culture, upbringing and environmental conditions will wire the frontal lobe in a unique pattern that determines individual's response to extreme stress,"
>
> -Dr. Kenneth Kamler, author of *Surviving the Extremes*.

The methods you use may be very personal discoveries. Generic in concept. Personal in execution. In the end, my friend? I want you to breathe the best breath of all, that sigh of relief when it's all really over and you are still in one "piece" and in one "peace."

Oh, and hey, skip the chewing gum. One more mouth-windpipe related thing. There are histories of people sucking gum down their windpipe when shocked, surprised or hit, lodging in and killing the chewer. While training? Good idea NOT to chew gum. Plus, gum has all kinds of weird and bad chemicals in it.

After the event? Get your act together! Breathe! Get rid of the hand quivers. Drink fluids too if you can. You might have an adrenaline dump to filter through.

HOW Question 6: How many counters are there? I start off every seminar with the disclaimer-warning, "Everything we do will have a counter. Surely more than one." And I add, "If you want to see a new counter to a move, bring completely untrained people in from the street and try it on them. You'll probably see something new."

This brings a smile from a veteran and a dropped jaw from the rookie. Rookies expect magic bullets and the magic tactic that always works. There are some high percentage successful ones, yes, but nothing is as close to a magic bullet as…well…a bullet, when it comes to counters.

On face value, a so-called "tactic," (military and police favored term) or "technique" (martial arts favored term) is a step or a series of steps to accomplish some level of diminishment of, or victory over, your opponent. And, on this face value, you might really favor a certain one or two because it seems easy and successful to you, based on who you are mentally and physically, along with your skills and expertise. Despite the user's varied specs, here are two choices:

   Type 1: Natural and Reflexive Counters.
   Type 2: Trained Counters.

Type 1: Natural and reflexive counters. Need we define these natural movements? For one example, if you feel you are falling or being taken down, and if you have the time, you usually step in the direction of the fall to counter the fall. A natural response. Or another example, a shoulder shrug or a rising arm are very natural ways people protect their heads. Obviously, the natural and reflexive counters are your worst problem. Everybody does them thoughtlessly. Reflexively. Most of the population is untrained and will react to you in these spontaneous, natural manners. They will probably respond in unusual, non-classic, off-the-chart counters.

Type 2: Trained Counters. Trained counters are different. They may be efficient responses that aren't necessarily so instinctive or intuitive, but rather learned, smart, and effective. In some cases, these trained counters when first learned even seem like foreign or strange movement. Counter-intuitive.

For example, if you are caught in an ambush firefight, one major counter is to charge the initial ambush while firing. This sounds crazy, but this is a trained response for several good reasons and vital when solving the common military rat-trap called a bigger secondary, planned ambush because there is another ambush set up for the common escape route. You are usually better attacking the first line.

Counter ratio. How many trained and untrained, natural counters are there versus a particular move? An enlightened study is required. This means getting with various experts and grilling them subject by subject, tactic by tactic. If your favorite tactic has eight easy, reflexive counters and five trained counters also, that is a bit high counter-ratio. And maybe that one move shouldn't be in your top 10 favs!

Also, the good news is when working on these lists with research and development, you are processing a lot of material, interacting with experts, crossing "borders," and becoming quite savvy about tactics, counters, and evaluation-evolution. This type of pro-and-con testing makes for a broad and unbiased spectrum of hand, stick, knife, and gun, crime and war survival knowledge.

We have already discussed in the WHEN Questions, the big "Before, During and After. One of its diverse connections is there are really 3 times for any counter: early phase counters, mid-phase counters, and late-phase counters.

1. Early phase counters. The enemy is about to start, or starting his attack.
2. Mid-phase counters. The enemy is in full action.
3. Late-phase counters. The enemy is starting to finish or finished his action.

Not all attacks have late-phase counters. For example, when a solid choke is settled in, there might not be a chance for a late-phase counter.

For example, in the small steps of fighting, what are the weaknesses of methods and moves. Trained fighters fight trained. Untrained fighters fight differently, with responses spastically and unusual compared to the trained people.

Pick a move. When you learn any move also learn the counters. You will do the move better. Make the lists. Start investigating.

HOW Question 7: How Do Fights Physically Start?
Often Two Basic Ways. I would like to identify two general types of argument and physical attack "starters." As we have established, whether in business, marriage, child-rearing, gang fights, or war, you are either interviewing, being surprised (ambushed), or have a few seconds, or longer, to assess the situation and take action.

In 1973, Sir Robert Ardrey wrote a groundbreaking book, *The Territorial Imperative,* which scientifically sought to prove that "everyone needs some space," particularly a primal space I and around them. There is discomfort, fear, or anger when this personal space is violated. His territorial ideas extended out to homes, states, and even countries.

This concept is crucial to understand, especially in training methods. A common aggressor may initiate a confrontation using two common fight starters:
1. The Stand-Off Confrontation.
2. The Mad-Rush Attack Confrontation.

1. The Stand-Off Confrontation. This is Collision 1 of the 6. This occurs when an aggressor stands before you, engaging in a confrontation "routine," which may include staring, bullying threats, yelling, finger-pointing, or even chest-bumping. You recognize the bully, the instigator, the pusher, and the sucker-puncher. This person deliberately invades your personal space, staying uncomfortably close to intimidating you.

If you attempt to create distance, they maneuver to maintain their dominance, often leading to sucker punches and strikes from non-traditional fighting stances. Some instructors may claim that allowing an aggressor to get too close is a mistake, but real-life situations can unfold rapidly, even for the most experienced individuals.

Modern police instructors advise officers to maintain at least two giant steps and a ½ lunge away from a suspicious person. However, even this distance can be breached in a split second. This scenario is often referred to as a "showdown" or "stand-off," known as Collision 1 of the Collision 6, when no physical contact has yet been made. During this phase, an opponent may stalk or attempt to corner you. Your best strategy is to keep moving and, if possible, place an obstacle like furniture or a car, whatever, in the way to you.

2. The Mad-Rush Attack Confrontation. I coined this term in the 1990s for our training programs and created instructional videos on the subject. This attack occurs when an aggressor starts at a slight distance, perhaps displaying a mean expression, taking a deep breath, or roaring before charging forward like a madman. Road rage incidents provide a clear example of this behavior, the aggressor has time and space to generate momentum for an all-out charge.

For both of these attack scenarios, I have developed numerous hand, stick, knife, and gun training drills, which are detailed in my other instructional materials and taught in seminars.

How Question 9: How complicated should fighting be? Dismissing the KISS Method. The pursuit of new knowledge and ideas is an ongoing endeavor. It reminds me of an old phrase I never liked: "Keep it simple, stupid." This is a shallow and, frankly, stupid motto. The K.I.S.S. method implies: "I am stupid, you are stupid, and we shall remain stupid."

Einstein offered a more thoughtful perspective: "Keep it simple, but not too simple." He understood that simplicity is relative, it varies depending on the person and situation. What is complex for one may be simple for another, and what is too simple for some may be just right for others. This presents a challenge for teachers and coaches: they must allow advanced students to think and train at higher levels to reach their own understanding of "simplicity." Leonardo da Vinci once said: "Simplicity is the greatest sophistication."

This wisdom underscores the importance of balancing simplicity with depth, allowing individuals to reach their full potential through tailored training and knowledge acquisition.

HOW Question 10: How to handle the argument confrontation. The splitting up of topics, based on the classification of questions in this book will always haunt me. I hope it doesn't bother you. We started out with verbal confrontations and arguments, and here in the How chapter we pick up some "hows" to handle them.

Indeed Career Guide has a nice kickoff to the subject of argument confrontation...

> "While many people associate the term, 'argument,' with a negative connotation, arguments are actually a regular part of the workplace and other areas of a person's life. Arguments are used to negotiate as well as to

determine the best solution for a particular issue. They can also be used to determine the extent of truth there is in a certain claim or hypothesis. There are several different types of arguments, and each type is used in different scenarios. Being skilled at arguing requires excellent communication and logic skills. These skills will also help you decide which type of argument is most fitting for the situation."

There are a few primary reasons why an argument may occur. These reasons include:
- To solve a problem or make a judgment.
- To defend or explain an action or stance.
- To communicate your point of view anyway of thinking to a person or group." - Indeed Career Guide

Since the quote mentions several different types of arguments, it behooves us to list them here, collected for your perusal now, and later for your bookshelf reference. Keep in mind several sources claim there is anywhere from 3 to 12 (or more) types.
- Deductive: A valid deductive argument is one where the conclusion is a necessity if the premises are true.
- Inductive: An inductive argument uses observations about the past to draw conclusions about the future.
- Abductive: An abductive argument uses a few relevant facts to draw a conclusion about what could explain them.

- Analogical: An analogical argument is based on the idea that if two things are similar, what is true of one is likely to be true of the other.
- Fallacious: A fallacious argument appears legitimate but is not.
- Causal: A causal argument is used to persuade someone that one thing caused another.
- Rebuttal: A rebuttal argument is used to refute an idea or belief.
- Proposal: A proposal argument proposes a solution to an issue. Evaluation: An evaluation argument determines if something is "good" or "bad."
- Narrative: A narrative argument tells a story to illustrate a point related to the argument.
- Toulmin: A Toulmin argument assembles evidence in support of claims.
- Rogerian: A Rogerian argument determines the best solution to an issue based on the needs of all parties.   - Professor Google

What if you are verbally ambushed, or testy subjects start up conversations. In his book How to Argue, Jonathan Herring does a good job outlining positive ways. He suggests this quick, wise outline. "They needn't be about shouting or imposing your will on someone. A good argument shouldn't involve screaming, squabbling or fistfights, even though too often it does. Shouting matches are rarely beneficial to anyone." Here are Herring's ten golden rules of argument.

- 1: Be Prepared. Make sure you know the essential points you want to make. Research the facts you need to convince your opponent. Before starting an argument, think carefully about what it is you are arguing about and what it is you want. This may sound obvious. But it's critically important. What do you really want from this argument? Do you want the other person to just understand your point of view? Or are you seeking a tangible result? If it's a tangible result, you must ask yourself whether this result you have in mind is realistic and whether it's obtainable. If it's not realistic or obtainable, then a verbal battle might damage a valuable relationship.
- 2: Learn When to Argue and When to Walk Away. Think carefully before you start to Argue. Is this the time? Is this the place? I'm sure you've had an argument before and later regretted it. Maybe you snapped at your spouse over dirty dishes after a long, stressful day at work, only to realize your frustration had nothing to do with housework. Learning when to engage in an argument and when to walk away is a vital skill. Most of the time, you're better off simply saying "you might be right" and letting it go. Why bother arguing? Is it really worth it in the big picture?

- 3: It's not WHAT you say but HOW you say it. Spend time thinking about how to present your argument. Body language, choice of words and manner of speaking all effect how your argument will come across. My mom told me this on repeat when I was a teenager. Be mindful not only of your words but your tone. People perceive threats unconsciously and respond accordingly. The best way to change minds is to be likeable - and most of that comes from how you say things.
- 4: Listen Again. Listen carefully to what the other person is saying. Watch their body language, listen for the meaning behind their words. You should spend more listening than talking. Aim for listening for 75 percent of the conversation and giving your own arguments 25 percent. And you're not listening if you're thinking about what to say next. This is often where a lot of arguments, and discussions for that matter, veer off course.

    If you're not listening to the other person and addressing their statements, you'll just keep making your same points over and over. The other person won't suddenly agree with those, and the argument quickly becomes frustrating. Imagine a couple arguing about household chores. If one partner keeps insisting, "You never help with the dishes," without hearing the other's explanation about their long work hours, they'll just go in circles, growing increasingly frustrated.

- 5: Excel at responding to arguments. Think carefully about what arguments the other person will listen to. What are their preconceptions? Which kinds of arguments do they find convincing. There are three main ways to respond to an argument:
    1. Challenge the facts the other person is using.
    2. Challenge their conclusions.
    3. Accept the point, but argue the weight of that point (i.e., other points should be considered above this one.)

- 6: Watch for crafty debate tricks. Arguments are not always as good as they first appear. Be wary of your opponent's use of statistics. Keep alert for distraction techniques such as personal attacks and red herrings. Look out for concealed questions and false choices.
- 7: Develop the skills of arguing in public. Keep it simple and clear. Be brief and don't rush.
- 8: Be able to argue in writing. Always choose clarity over pomposity. Be short, sharp, and to the point, using language that is easily understood. In many ways, thinking is writing.
- 9: Be great at resolving deadlock. Be creative in finding ways out of an argument that's going nowhere. Is it time to look at the issue from another angle? Are there ways of putting pressure on so that the other person has to agree with you? Is a compromise possible?

- 10: Maintain relationships. This is absolutely key. What do you want from this argument? Humiliating, embarrassing or aggravating your opponent might make you feel good at the time, but you might have many lonely days to rue your mistake. Find a result that works for both of you. You need to move forward. Then you will be able to argue another day. Another approach to end arguments is to ask the other person to explain their thinking.
- *Note:* The book takes a much deeper dive into these golden rules. Thanks, Mr. Herring.

I would like to add these two very important points. One is do not escalate the argument. Two, show zero signs of facial, or one increase. Remain as aloof as possible. Some people feed on this, love these reactions. Don't give it to them.

HOW question 11: How do you detect and avoid an ambush? Citizen. Police. Military. An ambush is not a "fight." It's a surprise attack from a concealed position on someone who is moving or temporarily stopped. As we covered earlier, the attacker picks everything:
1. The target (you or your team).
2. The time.
3. The place.
4. The method.
5. That's why ambushes are so deadly, whether it's a street robbery, a cartel roadblock, a police

car getting lit up at an intersection, or a military patrol walking into a kill zone.

Criminals ambush victims by using surprise, concealment, and sudden, overwhelming force in planned (entrapment) or spontaneous attacks, often targeting vulnerable moments like traffic stops, quiet streets, or by luring people with fake emergencies, while exploiting predictable patterns to create a "kill zone". Common methods involve hidden attackers, disguised scenarios (fake calls/help), or capitalizing on distractions, turning ordinary situations (like a vehicle stop or answering the door) into sudden, violent assaults.

Criminals ambush police by luring them into traps (entrapment) or exploiting sudden opportunities (spontaneous attacks), using surprise, concealment, and speed to attack with firearms or other weapons, often during routine calls like warrant service or traffic stops, by exploiting officer isolation, or creating hostile environments with pre-existing conflict. Methods include staged incidents, targeting officers alone in vehicles, or attacking during high-risk entries, aiming to overwhelm the officer before they can react.

Enemy soldiers ambush military forces by using concealment and surprise to unleash overwhelming, focused firepower from hidden positions, often targeting vulnerable choke points like trails or roads, using tactics like flanking, double envelopment, or mechanical ambushes (tripwires) to trap units in a "kill zone," disrupt their movement, and destroy or capture

them using overwhelming fire and rapid assaults. The goals are much the same across citizenry, crime, police work, and combat:
- Capture or kill people (or one VIP).
- Steal supplies, vehicles, or weapons.
- Block movement and choke off routes.
- Intimidate and demoralize.

The bad news is if you stumble blindly into a well-planned ambush and freeze on the spot, your odds are terrible. The good news? Ambushes follow patterns. The terrain, the route, the behavior of people and vehicles, and your own habits all leave clues. With some basic "protective intelligence" skills and a few simple crisis rehearsal plans and battle drills, you can avoid many ambushes and survive the ones you can't avoid.

Expert advice says step one: "Don't Be There When It Starts." The best way to survive an ambush is not to be on the "X" when it kicks off. The "X" is the spot where the enemy wants you to be when they attack. Could be:
- The intersection.
- The parking aisle.
- The doorway.
- The blind turn.
- The choke point.
- Below the high ground.
- One narrow hallway, one bridge, one stairwell, one lane funneling into a stoplight. Where do they have cover and concealment and I don't? Corners, parked cars, dense bushes, dark

doorways, abandoned vehicles, buildings overlooking a road.
- Where am I forced to slow, stop, or bunch up? Intersections, gates, parking garage ramps, traffic jams, toll booths, checkpoints, blind corners, speed bumps. Parking structures and far corners of lots. Walkways between buildings. Stairwells, elevators, side alleys Transitions between public streets and your car or front yard or the borders or your job and life..
- Those are chokepoints. That's where ambushers love to put "the X. You find the "X" by doing route analysis. Route Analysis (for everyone), whether you're a citizen just walking from your car to a store, a cop driving a beat, or a soldier moving a patrol, you should look at your routes through the eyes of an attacker. Ask yourself more of the WHERE questions. Where can someone control my movement?
- The X is endlessly situational.

You don't have to be paranoid. You just need to pay more attention in fringe and chokepoint zones and avoid handing attackers perfect setups. If trouble is possible, don't be time-and-place predictable.
- Vary departure times. Vary routes.
- Don't loiter on predictable X's (same pump at same gas station, same parking spot by the dark corner, etc.)
- Walk and drive your routes with an ambush eye.

- Reverse engineer your attack. Who, what, where, when, how and why would you attack yourself.

No ambush magically appears out of thin air. Somebody has to identify you as a victim, or a target. That means eyes on you at some point. That's where surveillance detection and attack recognition come in. What surveillance (or stalking looks like)

- In civilian crime, it may be one or two people. In higher-level attacks, it's a team. Either way, look for correlation.
- People or vehicles that move when you move.
- The same face popping up along your route.
- Someone loitering with a good view of your comings and goings (park, coffee shop, parked car, bus stop, etc.)
- Watch for behaviors, not costumes.
- Someone who's "busy" but keeps sneaking looks at you.
- Poorly faked normal behavior: reading a newspaper that never turns pages, pretending to talk on the phone but staring holes in you.
- Sudden interest when you move toward a chokepoint or chokepoint-fringe area.
- If you see correlation, don't shrug it off. You don't have to start a fight—but you do need to change something.

HOW Question 11: How do you mentally survive critical incidents? In a previous discussion, the topic of PTSD and critical incidents was explored. With each critical incident, the spectrum of post-traumatic stress looms over us all, yet the conversation surrounding it often focuses solely on the negative.

During a recent military training session I attended, a Special Forces veteran and presenter expressed a thought-provoking perspective, he wished the "T" in PTSD, which stands for "Traumatic," could be removed. He explained that if someone experienced a serious, stressful event and had no subsequent stress response, he would be more concerned about them than someone who did. His baseline concept was "Post-Stress Syndrome," which, in his view, branches into both positive and negative outcomes, including personal growth.

> "Not everyone who suffers trauma experiences post-traumatic growth, but or those who do, the changes can be lifelong. Although the exact number is unknown, researchers estimate that half to two-thirds of trauma survivors may experience post-traumatic growth."
>
> - Psychology Today Staff

Those in military and law enforcement circles understand that some individuals, under specific circumstances, experience severe post-stress reactions, ranging from treatable to debilitating. However, others, exposed to the same circumstances, still experience post-stress yet emerge stronger from the ordeal.

The outcomes vary, good, bad, negative, and positive. And yet, the positive side of this spectrum is rarely discussed. The media and internet are saturated with stories of trauma victims and their struggles, while accounts of resilience and growth remain largely unpublicized. A psychotherapist once told me that this positive response is often referred to as "resilience."

However, to me, that term feels incomplete, it fails to fully capture the dual nature of post-traumatic experiences. Other professionals use a more fitting term: Post-Traumatic Growth (PTG).

Yet, unlike PTSD, PTG is not a widely recognized concept. It is not a cause. It is a benefit with no victims. In a recent lecture to a large audience, I asked if anyone had heard of post-traumatic growth. Only one person, a psychologist, had.

Researchers went to West Point and asked how many were aware of post-traumatic stress disorders. The answer was 97%. How many were aware of post traumatic growth? Only 10%. Awareness and lack thereof are crucial when considering future outcomes.

Similarly, author Nassim Nicholas Taleb, in his book *Antifragile: Things That Gain From Disorder*, noted that there is no singular, non-hyphenated word for the opposite of fragile. This highlights the linguistic and conceptual gap in how we discuss strength emerging from adversity.

In the U.S. military and the Veterans Administration, PTSD is rated on a scale from 0 to 100, with 100 indicating the most severe symptoms. The most common rating is 70%, though ratings of 50%, 30%, and 10% also exist. The VA evaluates PTSD based on how it affects a veteran's ability to function in social and occupational settings. On the positive side is post traumatic growth (PTG):

"Many people increase in personal strength, appreciation of life, emotional intimacy with partners and family, creativity, sense of spirituality, and life possibilities following traumatic events and a more open attitude towards others, a greater appreciation of life and the discovering of new possibilities.

PTG may feature positive changes in self-perception, interpersonal relationships and philosophy of life, leading to increased self-awareness and self-confidence."

- Tedeschi & Calhoun, Measuring the Positive Legacy of Trauma, 1996

Let's break these out in a digestible, important list because followers of Tedeschi and Calhoun call these results the "5 Pillars of Post Traumatic Growth." The key word right now is "results."

- Pillar 1: Appreciation of life.
- Pillar 2: Emotional intimacy with friends and family.
- Pillar 3: Creativity.
- Pillar 4: Sense of spirituality.
- Pillar 5: Life possibilities.

Nice results, but how do people get these results? What kinds of people get these positive results? How? Tedeschi & Calhoun refer to people having "ego strength." Ego has become sort of a negative dirty word in the woke-world's vernacular. We hear "Oh, he has such a big ego!"

Experts do make the distinction between "ego-need," and "ego strength." I think simple common sense tells us the differences, but of course psychologists must deep dive and dissect. That's okay, that's their job...

> "Ego strength is described as a personality characteristic that allows dealing with adversity, managing external stressors and internal distress, and bouncing back from setbacks."
>
> - ScienceDirect.com

So, do you have a Survivor Personality? The ego strength? What skills do you have and what do you need to improve? Al Siebert, PhD in his books, *The Survivor Personality and The Resiliency Advantage,* offers a lot of tips. Here are a few I culled.

- Tip 1. Ask questions. Developing your curiosity to new situations increases your ability to understand the big picture of what is happening.
- Tip 2. Increase your emotional and mental flexibility. Tap into your biphasic (2 but maybe more of your "hats") personality traits in different situations. Survivors are creative, flexible and adapt to their circumstances.
- Tip 3. Accept change with uncertainty is a way of life going forward. Change is going to happen in life. Commit to hunting the good in every situation. If you look for opportunities in hardship, you will find them.
- Tip 4. Learn from all experiences. Failures are different from mistakes. Failure is when you

stop trying. Mistakes are lessons learned from how not to do something.
- Tip 5. Make time to observe and reflect. Take time to scan your surroundings and observe what is happening around you. Set your emotion aside and reflect on the facts. Do you see red flags or warning clues of pending troubles?
- Tip 6: Follow the survive-and-thrive sequence. Balance your emotions, adapt-cope in your immediate situation, thrive by learning and improving the situation for those involved, and then find the gift in your hardship. You can transform hardship into growth and become a survivor. With each adversity, your toolbox of resiliency skills grows, and you become mentally and emotionally stronger. A difficulty that almost breaks your spirit can become one of the best things that ever happened to you.

There are other helpful books and lists. Early on, I promised you that I would not pretend to be a psychologist, just a referrer, but being actively involved in doing this for years now, I have drawn a few of my own generic conclusions I wish you might consider.

I believe acquiring "ego strength" also requires these following building blocks, constructed with the help of the Ws and H questions of course, a few of my extra researched notes...
- Experts say one needs a higher "intellectual I.Q.," and a higher "emotional I.Q." to process what is necessary. These are inherent qualities

that not everyone has. Reminds me of my comment that "not all people can play professional football." Some can, some can't. Some have. Some don't. Some people can do certain things and some people can't do these things. Some have, some have not.

- Who inspires you? In an earlier essay I asked, "Who are your heroes?" Trying to be like a positive hero is inspirational. Asking that question when faced with problems. "What would _____ do now?" Where do you find inspiration?
- What is your belief system? For decades, experts have preached that movers, shakers and survivors need a belief system, or several systems. It could be a religion. It could be Jesus. It could be patriotism. It could be...revenge even. What keeps you going?
- The Big Comebacks, a personal note. A hero's journey. I will confess as to what has helped me through death, love, the military and policing. Ever since I was a young kid, Like I said, but not just being "hero-hero." I also became obsessed with heroes and their *journeys*. Their comebacks. Stories. From comics, to TV shows, to movies, to books, the Bible. Revival. Redemption. Resurrection, The hero's journey. Homer's Odyssey. The comebacks. The diversity. Adversity. The comebacks require mental and physical strength. The hero...comes back.

- I could be crushed by something tomorrow. Overwhelmed. But so far? So good. I can only say, be the hero in your story.

We must remember that people are different. There are those who survive certain things and situations better than others. Still, they might collapse into negative trauma from other specific situations and circumstances. Life is a genetic riddle, a flesh and blood Rubic's Cube that can get twisted way out of shape by the hand of time, chaos, bad luck and evil.

HOW Question 12: How small is your mind? We all know how tribal humans are. You belong to groups, have jobs, churches, teams, neighborhoods and political parties. Groups. "Group think." It's just being human. We begin to take on characteristics and ideas of these partitioned groups. Totally natural. So, the question remains, how can you evaluate your free-thinking mind? How insulated, how much of a "mental bubble" insulates, protects, limits your thoughts? Keep in mind while this is a natural trait of the human race with positive traits, but it can be tunnel vision.

How about the classic tale from India of blind men touching the elephant. The Indian parable of The Blind Men and the Elephant dates back to around 500 B.C., appearing in the Buddhist text Tittha Sutta. The story has been adapted in many forms and is part of many religious traditions, including Hindu, Jain, and Buddhist texts. Perhaps the idea of the 1830s origin comes from the American poet John Godfrey Saxe (1816-1887)

who retold and popularized the story in his poem, The Blind Men and the Elephant. It basically goes like this:

"A group of blind men heard that a strange animal, called an elephant, had been brought to the town, but none of them were aware of its shape and form.

Out of curiosity, they said: "We must inspect and know it by touch, of which we are capable" So, they sought it out, and when they found it they groped about it. The first person, whose hand landed on the trunk, said, "This being is like a thick snake."

For another one whose hand reached its ear, it seemed like a kind of fan. As for another person, whose hand was upon its leg, said, the elephant is a pillar like a tree-trunk. The blind man who placed his hand upon its side said the elephant, "is a wall."

Another who felt its tail, described the elephant is like a rope. The last felt its tusk, stating the elephant is that which is hard, smooth and like a spear."

Saxe's poem ends with the explanation that the elephant is a metaphor for perception and thinking. No one has full experience. The poem warns readers that preconceived notions, and small perceptions can lead to misinterpretation. Inbreeding of the mind and body usually has negative results. Even extinction!

I am not trying to suggest that groups, jobs, churches, teams, neighborhoods and political parties are inherently bad for the brain. No. Just think about this. Make educated decisions about your commitments. Keep inquisitive. Keep learning. Asking.

How Question 13: How do I make my office workplace safer? As a police detective in the Army and in Texas, I had a desk in an open bay office area of desks in several buildings as the police agencies moved locations. I never had any family pictures on or around my desk, as other did, which that (and a few other things) drove me crazy. There were always a broad cast of characters, witnesses and suspects, seated at my desk, and other desks through time, some people very dangerous and crazy. They should not be gazing at photos of my wife and kids.

"Other things?" Here is a list of safety measures to make your office bays and work areas safer. No matter what your business, you should counter workplace violence from co-workers, customers, visitors and...suspects. Aside from organizing the layout to prevent people from slipping and falling, and worrying about clean air...

- Remove objects that could be used as a weapon.
- Rearrange your office so that you and yours have equal and unimpeded egress. Have an escape plan.
- Inner and outer office communication should allow you to communicate with others very quickly and easily, as well as alert police if they are needed.

- Unnecessary possessions should never make it into your private and restrictive areas.
- Train your clerical staff to be able to verbally calm people down and recognize possible potential dangers. (Tell them to read this book!) If even Starbucks does this, why doesn't your job too? Starbucks still trains employees to calm down hostile customers, requiring de-escalation tactics for situations like people refusing to buy items or leave the restroom, using specific techniques like active listening. They use LATTE: Listen, Apologize, Take action, Thank, Ensure satisfaction), making eye contact, and maintaining a calm tone to handle conflicts, though some baristas worry it isn't always enough.
- Install unobtrusive, yet protective barriers where they need to be.
- Try to never work alone.
- No family photos lingering about.
- Evaluate where you park your cars at work.
- Do not put magazines with your home address in the waiting rooms.
- Hire security if needed. Even some police agencies and militaries have hired specialized security guards.
- Gossip? Yes, keep an ear to the ground. Personal and/or domestic problems can lead to workplace violence, even murders.

You will have to customize a working safety list for where you work. These are just some general ideas. I could tell you so many stories about incidents and crimes surrounding this topic. Here's one, even before I entered police work...

Keep an eye on Jerry. While attending college in Dallas, I worked in a metal factory where I welded, shaped and made all sorts of metal products. There was an employee there who was very adept at working a complicated metal shaping machine. It was heavy, hands-on labor. He was a white guy, a loner, always dirty, about 50 years old with crazy unkempt hair. He was as crazy as his hair.

The place was very noisy but on occasion, he would curse and shriek like a wild man above the din. Shocking to me at first, I got used to it along with all the other busy bees at work. But also, about every few weeks, something would go wrong with his machinery (I presume) and he would throw a hammer and scream-roar. At no one. He would just heave it across the open factory floor like Thor killing a monster. It would clang and bang off the piles of metal and machinery.

As far as I knew, no one ever got hit? The floor supervisors would collect the hammer and cautiously approach him, returning the tool. The gossip on the floor was, "Keep on eye on Jerry."

I eventually joined the Army and as far as I could tell, he never killed anybody. That was the early 1970s. I can only imagine what would happen to such a hammer-thrower in todays' timid, nanny world. Anyway, the incredible diversity of workplace problems is near infinity.

How Question 16: How can I de-escalate an angry attacker? Earlier in this book, I promised to cover some information on this tricky topic. De-escalation is, at best, a "How-to" question. I have confronted numerous angry, even crazed, suicidal people. I've attended a number of these courses in my policing years. And frankly, it's incredibly difficult to de-escalate some people for the reasons I will now outline.

1. Who are you trying to de-escalate? Each situation requires a different approach. Who you are dealing with shapes your approach. Are you dealing with:
    - A madman? (Good luck.)
    - A bully? (He thrives on conflict.)
    - A mugger? (He's a no-nonsense, hit-and-run thug.) A terrorist? (Good luck again!)
    - An angry spouse? (Good luck again.)

- An upset customer? (They want a business deal or settlement. Maybe some...revenge.)
- A hostage taker? (They want escape and/or money.
- A suicidal individual?
- A workplace conflict between co-workers?
- A confrontation in a high-stakes environment, such as some sort of a battlefield? Someone else entirely?

2. What is happening? Each situation demands a unique response. What do they want? What will you say or do to influence the outcome?

3. Where is this happening?. The location can heavily influence the dynamics of a confrontation.
- His, her, their or someone else's home.
- His, her, their or someone else's work.
- His, her, their or someone else's school.
    - A hospital?
    - A battlefield?
- When is this happening? Timing plays a critical role.
    - A weekday or weekend?
    - A holiday or significant date?
    - An overt or covert deadline at play?
- How is it unfolding?
    - How did this escalate?
    - How is the situation progressing?

- How many possible endings are there?

4: Why is this happening? Understanding the root cause of the conflict is essential for resolution.

The tip of the iceberg. De-escalation is far more complex than most people assume. Over the years, I've talked people out of suicides, convinced them to surrender, persuaded them to drop their weapons, and even freed hostages. While SWAT teams now have trained negotiators, it's often the patrol officers or detectives who arrive first and must handle the situation before backup arrives. Citizens are often the first responders.

I recall once talking a suicidal man off a three-story apartment building roof. I told him that, according to my police academy training, it takes at least a four-story jump to ensure death, anything less would probably leave him paralyzed. This was a complete lie, but it worked. He climbed down the fire truck ladder with me. That kind of improvisation isn't something you'll find in negotiation training courses.

Many de-escalation courses focus on mediating workplace conflicts between co-workers or customers. They emphasize understanding tones and sympathy, techniques that often fail in high-stakes situations. In reality, de-escalation requires more than just a calm demeanor. It demands charisma, quick thinking, and sometimes even a strong physical presence. It is all VERY situational. Not everyone can be an artist, but there is an art to de-escalation.

The power of "going off script". In this book, I've discussed the "scripts of life" and how people in various situations often repeat the same predictable dialogue. But sometimes, an unexpected remark can completely shift the dynamic. I've used this technique, and I teach others to do the same. I call it "going off script." I've received messages from as far as Australia, thanking me for this approach. And now I will give you professional formal de-escalation lesson plan from my attendance notes...

- Step 1: De-Escalation – Start with yourself. Before you can calm anyone else down, you've got to de-escalate yourself first. If you walk into a confrontation already running hot, you'll only feed the fire. Recognize the Fire Inside. Notice your state, anger, fear, frustration. Name it, don't hide it. If you don't spot it, it will drive you.
- Step 2: Separate You from the Emotion. You are not your anger. You are not your fear. Those are just chemical storms. Step back from them and take the driver's seat.
- Step 3. Control What You Can. You can't control the other guy. You can control your posture, your tone, your words, your distance. Stay in charge of those.
- Step 4. Expect Trouble. Be Ready Anyway. Picture the worst-case outcome. That way, if it happens, you're not shocked, you're prepared. That steadiness is its own weapon.
- Step 5. Act Like the Adult in the Room. You are the grown-up, the professional, the protector.

Don't let a moment of rage wreck your reputation, your job, or your freedom.

Some De-Escalation Techniques. Once your head is clear, you can work on the other person. Here's how the experts suggest:
- Listen Hard. Let them talk. Really listen. Don't cut them off, don't roll your eyes. People may calm down faster when they feel heard.
- Show Respect Without Surrender. Say those classic things like:
  > "I hear what you're saying."
  > "I get why you're upset." That doesn't mean you agree. It just lowers the heat.
- Find the Bridge. Look for shared ground: safety, fairness, respect. Even a thin bridge is better than a canyon between you
- Control Your Body First. Hands visible. Shoulders down. Stand tall but not aggressive. A calm body broadcasts a calm mind, even if you're ready to move in an instant.
- Keep the Voice Low and Slow. The louder and faster they get, the lower and slower you go. This contrast often drags their tempo back down.
- Offer Logical "Outs." Give the other person a face-saving way to back down. Nobody wants to feel cornered. Examples:
  > "We can step aside and talk."
  > "Let's solve this without getting anyone in more trouble."

- Know When Talk Time Is Over. Sometimes words won't work. If safety collapses, be ready to move. De-escalation is a skill, not a magic spell.
- Distance is de-escalation. If you can, put more space between you and them. Obstacles are de-escalation. A table, a shopping cart, a car, anything between you buys time. Escape is de-escalation. The best "win" may be walking away intact, if you are alone or can abandon the situation?

A great (and humorous) example of "going off script" is found in the old James Garner films like *Support Your Local Sheriff*. Garner constantly surprises confrontational characters with unpredictable remarks. Another great example is Raylon Givens in the TV show *Justified*, as he dodges threats. Another great verbal side-stepper example is James Spader playing Reddington in *Blacklist*. Of course, none of us are as effortlessly cool as Garner, Givens or Spader, so we must develop our own engaging personas and clever retorts.

Effective de-escalation isn't just about words, it's about presence, adaptability, and knowing when to break the expected script.

How Summary: I think that the HOW questions are very important. Continue to investigate the small and big HOW questions and answers.

The HOW Question Review.

Who are you within this how category?
What are you within this how category?
Where are you within this how category?
When are you within this how category?
How are you within this how category?
Why are you within this how category?

Chapter 8: The Why Question and Confrontations

I usually mention in each seminar, this example:
> "When someone is on top of you, punching you in the face, you don't pause to ask yourself, 'Why is this poor gentleman so mean, so angry? Could Gestalt Therapy help him?"

No. In that moment, you're not asking why. But before and after such events, during preparation, analysis, or reflection, asking why holds significant value. It helps with criminal and enemy soldier profiling, forecasting threats, and examining cultural and societal dynamics. The question WHY is ultimately an academic pursuit.

You can ask anyone about anything: Why did you do this or that? Some may say, "I don't know." But others do know. It all comes down to motives, decisions made beforehand or in the heat of the moment. The study is academic and may have future crime problem solutions and predictions.

WHY Question 1: Why do people commit evil?
Explaining all the causes and motivations behind evil is nearly impossible. Some, like me, believe we live in a messy, fallen world full of tests, trials, and tribulations. Evil and tragedy are part of that testing. While we can point to psychological flaws and societal issues, the motives behind evil range from simple to as complex as the human mind itself.

According to Psychology Today, there are five major root causes of evil, factors-reasons that can drive even ordinary, well-intentioned people to commit harmful acts, whether alone or in combination:
- The desire for material gain.
- Threatened egotism.
- Idealism, the belief that noble ends justify violent means. (Thus, the saying: "The road to hell is paved with good intentions."
- The pursuit of sadistic pleasure, though research suggests only 4–6% of perpetrators actually enjoy inflicting harm.
- Unbalanced behavior, including anger, revenge, hatred, psychological trauma, selfishness, ignorance, destruction, and neglect.

Understanding why doesn't always prevent evil, but it helps us recognize patterns, anticipate dangers, and perhaps, in some cases, get to forecast and intervene before harm is done.

WHY Question 2: Two major geography-WHY questions. I bring this up again as a reminder in the later pages of this book. Those two questions might haunt your survival-fight and legal life that makes you a fugitive or puts you in jail and/or costs you and yours thousands. Maybe a million? Those who build cases in your defense and those who wish to destroy your cases, will torture you on these two whys'.
1. Why did you go there?
2. Why did you stay there?

You have to be prepared to logically explain why you had to hit, strike, stab or shoot someone.

The worst-case scenario. There's about to be fight. Or you are roughed up. You leave for your car. But you return with a knife or a gun from your car. Stab. Or bang. The legal world will want to know why you went back. Why did you stay? The legal world wants you to leave when you get to your car. You might have other reasons to return, like the safety of others still there. Sure. You'd better have good reasons to go back.

Now change the nouns. Car. House. Store. Restaurant. Campsite. Parking lot. Sports arena. Bar. Motel. ETC. Change the weapon. Hand. Stick. Knife. Gun. Chair. Hammer. Etc.

Let's go a little further, linguistically. Change the "adjectives" the descriptions of the nouns, like big gun, small gun, big guy, small guy, long walk back, etc. What about the verb? It's that actual verb that gets you in and out of trouble. Run, walk, shoot, stab, strike...

WHY Question 3: Why do some survive, and some don't? On my bookshelf beside me right now as I type, I have 11 survival books written by some terrific historians, psychologists, doctors, and military and police veterans on the subject of surviving. The classic list of conflicts are:
1. person versus person,
2. person versus self,
3. person versus nature,
4. person versus technology,
5. person versus supernatural,
6. person versus fate, and,
7. person versus society.

These experts are expected by their publishers to summarize their books with a list of "what qualities does it take to survive." I can tell you that a plethora of qualities were listed by all these experts. In some books the lists and qualities were very long. Some very short. Imagine that. A few experts just say "resilience," or resilience appeared in most books. Great, but what constitutes and develops resilience? What experts say is resilience:
- Resilience is the capacity to withstand or to recover quickly from difficulties. Toughness. Elasticity.
- The ability to bounce back from difficult experiences.
- The ability to maintain psychological well-being in the face of adversity.
- The ability to keep going, both physically and psychologically, after something bad happened.

- The process of successfully adapting to challenging life experiences.

How experts say resilience is developed.
- Resilience can be developed through learning physical and mental skills and practicing behaviors.
- Resilience can be developed through recognizing strengths that may not have been known before.
- Resilience can be developed through developing coping strategies.

And I would like to make a quick reference a very successful U.S. Army program called Master Resilience Training, (M.R.T.) pioneered by Dr. Martin Seligman, director of Positive Psychology at the University of Pennsylvania. Describing it here would be a rather book length endeavor, but it functionally implements many of the topics from above. Seek it out in your pursuit

Historically some have survived with traits like anger, revenge and vengeance. Like it or not, negative or not, they can still get and keep you moving and alive.

Please refer back to How Questions about posttraumatic growth for part of this answer. I saved this quote for this part:

"Learning about post traumatic growth, learning to ask how could these experiences serve us? And being pushed to own the experiences that we have been through and use them to fuel our future, proved a powerful tool in helping our individuals, teams and organization thrive, not in spite of stress but because of it."

- SEAL Commander Curt Cronin

WHY Question 4: Lines in the Sand. Why then should I fight? Lines in the sand. It might take a legitimate, historical chapter to record all the references of lines in the sand. There's even one in the Bible. Perhaps the most popular is the one drawn by Colonel William Travis at the Texas Alamo.

The story goes that during the siege of the Alamo, Colonel Travis, facing certain defeat, gathered his men and drew a line in the dirt with his sword. He then asked those who wished to fight to the death to cross the line, while those who wished to surrender could stay on the other side.

We are down to it now. One might say after reading this book, fighting is a lose-lose event. One might say - "I'll wind up in expensive trouble. Hurt, in jail, sued or dead."

Here in this very last question, the "why' question, I would like to make a statement about that. This is the sad story of humanity. Centuries worth. Lots of lines in the sand. Read history. Read the Bible.

I have extensively toured the castles and churches in Europe, took the walking tours, read the books about Medieval Europe and times like the Crusades. And I would like to leave you with a few words from The Fourth Crusade and The Sack of Constantinople by Jonathan Phillips to display this timelessness:

"The most pressing emotion for the crusaders, as the majority of soldiers throughout history have also experienced, was the fear of death and captivity. The brutality of the medieval battles would have been familiar in the West, but the hardships of a crusade, the distance, the disease, the climate, the unknown enemy and the food supply would hold additional terrors. As people have discovered following recent episodes such as Vietnam and the Gulf conflicts, it is easy to overlook the mixture of emotions generated in the aftermath of a crusade. For some the homecoming brought fame and joy. Men's achievements were enshrined in oral verse and literature." - Jonathan Phillips

For some fame and/or joy? Yet, for others fated to the mental and physical tragedies of dungeons, death and dismemberment. All one has to do is change some of the geographic nouns to consider the timelessness of it all.

Reader! Consider all these motivations before taking action in fights, crime and war. If someone demands your wallet, your car, your loved one, a friend or stranger, demands your house, your country? Where is your line in the sand?

Yes, it is easier and safer to turn over the monies, the treasures, the geography. But why? Why do it? Think about your life and that line in the sand. If someone invades your life, your country? Shreds up your Constitution. Wreaking havoc on decent people? Where is your line in the sand?

It is easy and wise for me in this book to warn you of the pitfalls, sure, but then, look at me. I confess! I joined the Army near the end of the Vietnam War. I was a police officer for three decades. So, who am I to judge on smart and dumb avoidance choices.

But I have a line in the sand. I am someone who was-is crazy enough, or patriotic enough, or angry enough, or foolish enough to stand up. To take risks in spite of all these possible troubles. To essentially... fight for something. For me, if there's any chance, I can thwart some of these acts? I will choose to bust these mother-fuckers up. Despite the risks? Do the right thing. Yes. But smartly.

Who am I, dare I say who should you be? A peacekeeper. This definition helps keep me in line. Think about it in crime and war. Keeping the peace. Blessed are those. When necessary? Force is necessary. Hand or stick or knife or gun necessary.

I won't do something stupid on purpose. I have tried and will try to fight on as smart as I can. But I cannot face my reflection in the mirror, I can't disappoint myself if I don't stand against crime, up for some honor, the country and justice. Stand for the right reason. It's this big picture I write about, and what this entire book is all about. Forecasting, helping you establish your lines in the sand.

Who are you, again I ask. What's in your mirror? You know the gamble. Where are your lines? Most people haven't thought about any lines at all, and they are…ambushed. Set the smart lines in the sand. Think ahead. Gamble. Forecast smartly by living in and with the lines, founded by the answers from all the questions, the who, the what, the war, the when, the how and the why. You've investigated some serious answers here, WHERE are your lines in the sand?

WHY Summary. The why of things is the most academic question of them all. Never stop investigating why things happen and why people do things. You might never find all the answers, but you have to keep trying. This is how discoveries are made. We don't stop evolving behind the guy who stopped in front of you.

The Why Question Review.

Who are you within this why category?
What are you within this why category?
Where are you within this why category?
When are you within this why category?
How are you within this why category?
Why are you within this why category?

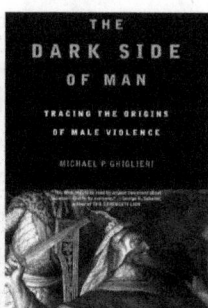

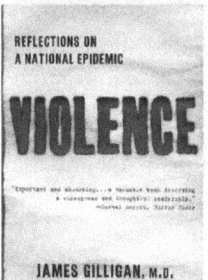
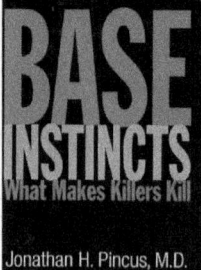
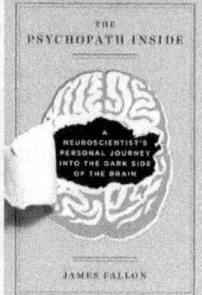

This is the end of this book, but the pursuit of Ws and H answers go on. Each week or month I still find new information to add. Now the pursuit is in your court. Hunt on.

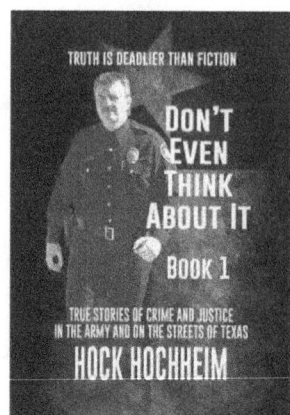

TRUE CRIME!
Book 1: The Patrol Years
Book 2: The Detective Years

Calls and Cases of a Veteran Lawman

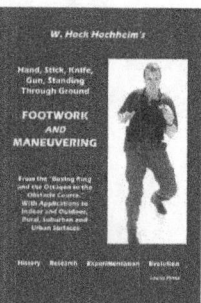

Free for Kindle Unlimited Members

These Kindle Ebooks are Only $10

You Can Still Get the Hardcovers and Oversized Paperbacks

The late 1890s. The early 1900s. A time just after the American gunfighter, and right before the noir detective. A time when men with a certain experience were called upon to solve difficult problems. Men like Johann Gunther, former military officer and war vet, ex-Texas lawman, spy and owner of a special firm called Remedies in Fort Worth, Texas. Due to Gunther's worldly experiences, he is often tasked with international adventures, bringing his special "code of the west," western hero ethics into the life and death of war and crime.

Book 1: Gunther! West of Medieval
Book 2: Guns of the China Alamo
Book 3: Last of the Gunslingers
Book 4: Riders of the Khyber Pass
Book 5: Rio Grande Black Magic
Book 6: The Horse Killers

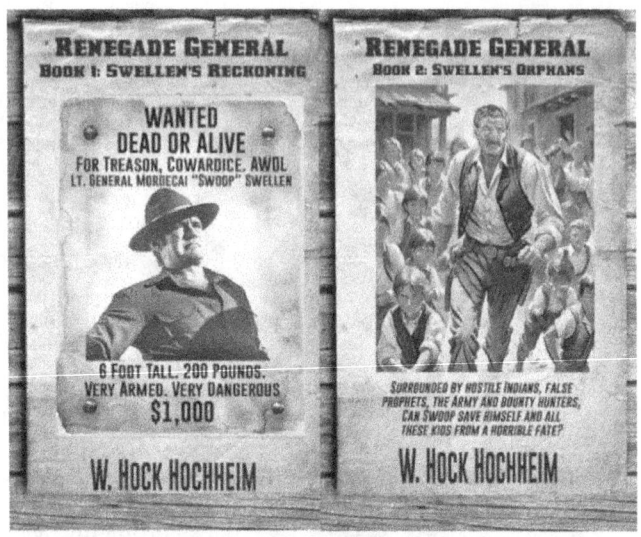

Falsely accused of treason, Swoop is back "and on the run" in this new eBook and paperback series, thanks to a passel of worldwide distributors such as Barnes and Nobles and Apple. If you like the classic westerns of old, like "Josey Wales," "Branded" and "One-Eyed Jacks," you will saddle up with Swoop Swellen as he travels the world to stay ahead of the U.S. Army, Wells Fargo and vicious bounty hunters, all the while falling into dangerous adventures and misadventures.

Book 1: Swellen's Reckoning
Book 2: Swellen's Orphans

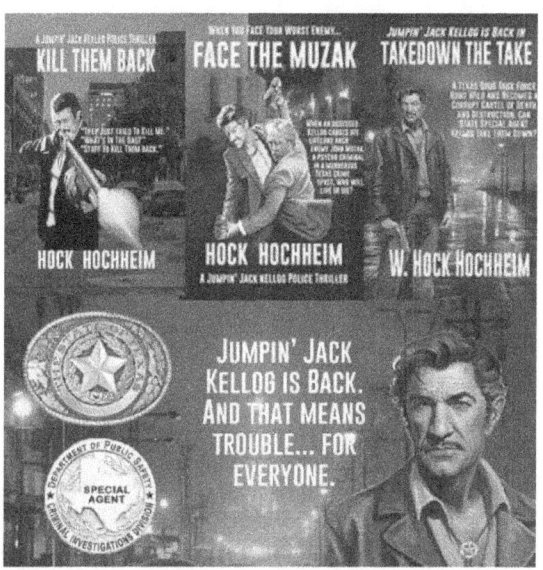

It's Texas in the 1980s and 90s, and his world is full of corrupt police, bad judges, seedy bail bondsmen, conniving complicit lawyers, the Cowboy Mafia, violent career criminals, cartels, desperate drug addicts, rapists, weaselly thieves and frightened snitches. he smells them out, stomps on them, puts them away or sometimes? Sometimes he has to kill them.

Some say that Jack Kellog is a "Texas Dirty Harry," or maybe call him a "rogue," but Jack is more complicated than that. He's a former Houston street cop, a Harris County city detective who almost died in the field, and was resurrected as a special agent in the Texas State Police Intelligence Division. There, he might work as a "lone wolf" or command a whole task force. The Governor knows him. The CIA know him. Mexican generals, the Cosa Nostra and the Cowboy Mafia know him. How about you? Do you know Jumpin' Jack Kellog?

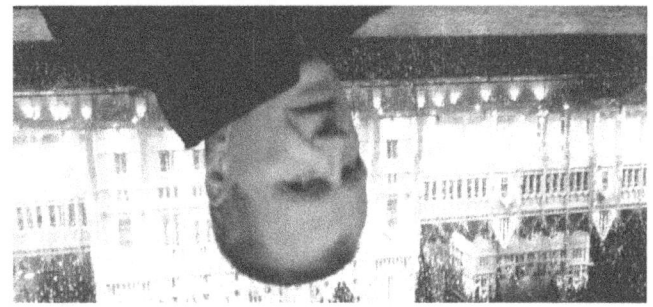

Find all of Hock's books on Amazon and Barnes and Nobles